Palgrave Studies in the History of Science and Technology

James Rodger Fleming (Colby College) and Roger D. Launius (National Air and Space Museum), Series Editors

This series presents original, high-quality, and accessible works at the cutting edge of scholarship within the history of science and technology. Books in the series aim to disseminate new knowledge and new perspectives about the history of science and technology, enhance and extend education, foster public understanding, and enrich cultural life. Collectively, these books will break down conventional lines of demarcation by incorporating historical perspectives into issues of current and ongoing concern, offering international and global perspectives on a variety of issues, and bridging the gap between historians and practicing scientists. In this way they advance scholarly conversation within and across traditional disciplines but also to help define new areas of intellectual endeavor.

Published by Palgrave Macmillan:

Continental Defense in the Eisenhower Era: Nuclear Antiaircraft Arms and the Cold War
By Christopher J. Bright

Confronting the Climate: British Airs and the Making of Environmental Medicine
By Vladimir Janković

Globalizing Polar Science: Reconsidering the International Polar and Geophysical Years
Edited by Roger D. Launius, James Rodger Fleming, and David H. DeVorkin

Eugenics and the Nature-Nurture Debate in the Twentieth Century
By Aaron Gillette

John F. Kennedy and the Race to the Moon
By John M. Logsdon

A Vision of Modern Science: John Tyndall and the Role of the Scientist in Victorian Culture
By Ursula DeYoung

Searching for Sasquatch: Crackpots, Eggheads, and Cryptozoology
By Brian Regal

Inventing the American Astronaut
By Matthew H. Hersch

The Nuclear Age in Popular Media: A Transnational History
Edited by Dick van Lente

Exploring the Solar System: The History and Science of Planetary Exploration
Edited by Roger D. Launius

The Sociable Sciences: Darwin and His Contemporaries in Chile
By Patience A. Schell

The First Atomic Age: Scientists, Radiations, and the American Public, 1895–1945
By Matthew Lavine

NASA in the World: Fifty Years of International Collaboration in Space
By John Krige, Angelina Long Callahan, and Ashok Maharaj

Empire and Science in the Making: Dutch Colonial Scholarship in Comparative Global Perspective
Edited by Peter Boomgaard

Anglo-American Connections in Japanese Chemistry: The Lab as Contact Zone
By Yoshiyuki Kikuchi

Eismitte in the Scientific Imagination: Knowledge and Politics at the Center of Greenland
By Janet Martin-Nielsen

Anglo-American Connections in Japanese Chemistry

The Lab as Contact Zone

Yoshiyuki Kikuchi

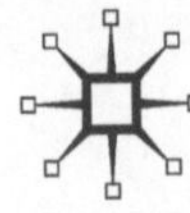

ANGLO-AMERICAN CONNECTIONS IN JAPANESE CHEMISTRY

First published in 2013 by
PALGRAVE MACMILLAN®
in the United States—a division of St. Martin's Press LLC,
175 Fifth Avenue, New York, NY 10010.

Where this book is distributed in the UK, Europe and the rest of the world, this is by Palgrave Macmillan, a division of Macmillan Publishers Limited, registered in England, company number 785998, of Houndmills, Basingstoke, Hampshire RG21 6XS.

Palgrave Macmillan is the global academic imprint of the above companies and has companies and representatives throughout the world.

Palgrave® and Macmillan® are registered trademarks in the United States, the United Kingdom, Europe and other countries.

ISBN: 978–0–230–11778–5

Library of Congress Cataloging-in-Publication Data is available from the Library of Congress.

A catalogue record of the book is available from the British Library.

Design by Newgen Knowledge Works (P) Ltd., Chennai, India.

First edition: December 2013

10 9 8 7 6 5 4 3 2 1

Contents

Figures

Tables

Preface and Acknowledgments

This book started its first life as a dissertation on Anglo-Japanese scholarly relations in chemistry submitted to the Open University in Milton Keynes, United Kingdom, in 2006, though research for the dissertation had started earlier while I was a graduate student at the University of Tokyo working with Takehiko Hashimoto. The OU, as it is affectionately called by students, faculty, staff, and locals, is no ordinary place to do a Ph.D. But there were many good sides to it. There was a research tradition, if not a school, in the history of chemistry created by the late Colin A. Russell in the 1980s and 1990s at the OU. He had already retired when I arrived in Milton Keynes in 2002, but I was fortunate enough to meet him and received invaluable advice from time to time, especially about the history of organic chemistry in which I had been ill-prepared.

Even more important was the intense and fruitful relationship with my supervisors. Gerrylynn K. Roberts and Ian E. Inkster spent countless hours discussing my topic, giving advice and counsel at crucial moments, and proofreading draft chapters in the final writing stage. The full extent of their generosity did not become clear to me until later when I heard about the experiences of other PhD students worldwide. I also had the good fortune of meeting a senior fellow student (*senpai* in Japanese) in sociology, Masaaki Morishita, who introduced me to the key concept of this book, the contact zone. I am proud of having been a student at one of the most forward-looking universities in the world and still cherish a fond memory of my experience there and my life in Milton Keynes.

The second life of this book, in hindsight, began when I moved to the United States to take up a postdoctoral position at the Chemical Heritage Foundation (CHF) in Philadelphia in 2008. I conducted key research there that would be necessary to add a crucial American element to this book. Hyungsub Choi, then at CHF, asked me to join his informal reading sessions (with mostly only two of us) in the history of science and technology in modern and contemporary Japan, giving me an opportunity to think about the wider relevance of my story beyond its immediate time.

However, it was after moving to the Massachusetts Institute of Technology (MIT) in Cambridge, Massachusetts, in 2009 that this book project started in earnest. David (Dave) Kaiser, my mentor there, encouraged me to really go for it, talked with me about reconceptualizing my project, discussed and proofread sample chapters, and helped me draft a book proposal—practically everything you would need to get a book project started. Dave also organized a stimulating discussion group on the history of modern physical sciences where earlier drafts of chapter 5 and the book proposal were read and discussed. This process continued after I moved to Harvard University in 2011 and there came under the care of two exceptional mentors, Shigehisa Kuriyama and Janet Browne.

A major part of the writing was carried out in 2012 and 2013 at the International Institute for Asian Studies (IIAS) in Leiden, Netherlands, directed by Philippe Peycam and managed by Willem Vogelsang. Originally trained as a historian of science, I was not sure at first what to expect in this institute, except that it is located in a beautiful historic university town with a tradition in Japanese studies. Now I can say with confidence that it was the best place to finish my book project. Particularly noteworthy is IIAS's firm commitment to colonial, postcolonial, and heritage studies and its vibrant research cluster for urban design called the Urban Knowledge Network Asia (UKNA). I did not become an expert in either field, but I received insights, stimulating discussion and conversation (distinctions between them were often unclear), and thought-provoking questions that sharpened my thinking and contextualized my project within broader Asian studies. Netherlands-based colleagues in the history of science and technology, especially Ernst Homburg (an old friend of mine), Lissa Roberts, Andreas Weber, and Martin Weiss kindly provided much-needed discussions and encouragement in my own area of specialization. Lissa also agreed to read an earlier draft of the introductory chapter and gave me invaluable advice on how to improve it.

My current employer, The Graduate University for Advanced Studies (Sokendai) and my new colleagues in its "Science and Society" teaching program—Mariko Hasegawa, Kohji Hirata, Kaori Iida, Kenji Itō, Kōichi Mikami, Hisashi Nakao, and Ryūma Shineha—supported me while I was adding finishing touches to the manuscript and kindly joined the stimulating discussion of a part of my book manuscript.

The above list of people and institutions that supported my project is far from complete. Yasu Furukawa convinced me that the history of chemistry was an exciting research field and initiated me into it in his seminar courses at Tokyo, and I am grateful for that. One of his suggested readings was *The Fontana (Norton) History of Chemistry*, written by William H. (Bill) Brock. It was by participating in its Japanese translation project, organized by Makoto Ohno, that I acquired the overall basic knowledge of this field. Later in England I had the privilege to get my dissertation examined by Bill himself together with Janet E. Hunter, a task they carried out with encouragement, constructive criticism, and helpful comments. Peter J. T. Morris generously shared his knowledge and enthusiasm about the history of chemistry, the chemical laboratory, and its design, which were both enlightening and infectious.

I am grateful to the following scholars (in alphabetical order) for taking part in the discussion of my research topic and for giving me invaluable advice: Juliana Adelman, James R. Bartholomew, Antonio García Belmar, Gwen Bennett, Harm Beukers, Gregory Bracken, Ronald Brashear, Robert F. Bud, Hasok Chang, H. Floris Cohen, Clara Cullen, Antonella Diana, Matthew D. Eddy, Graeme J.N. Gooday, Catherine E. Guise-Richardson, Jonathan Harwood, Hiro Hirai, Aya Homei, Shuntarō Itō, Chihyung Jeon, Jeffrey A. Johnson, Masanori Kaji, Robert H. Kargon, Masae Katō, Aarti Kawlra, Dong-won Kim, Stuart W. Leslie, Morris F. Low, Roy M. MacLeod, Ethan Mark, Ben Marsden, Hajime Mizoguchi, the late Minoru Nakano, Takuji Okamoto, Takeharu Ōkubo, Christina Pecchia, John Perkins, John V. Pickstone, Jennifer M. Rampling, the late Masanobu Sakanoue, Tom Schilling, Grace Y. Shen, Josep Simon, Matthew Sindell, Anna E. Simmons, David Roth Singerman, Pierre Teissier, Anke Timmermann, Tōgo Tsukahara, Carsten Timmermann, Masao Uchida, Rob van Veen, James R. Voelkel, Masanori Wada, Stephen Weininger, Benjamin Wilson, Michael Worboys, Masakatsu Yamazaki, and Kenji Yoshihara. In addition to giving me information and advice, Yasushi Kakihara, Mari Yamaguchi, Hideyuki Yoshimoto, and Toshifumi Yatsumimi helped me collect materials located in Japan.

At the publishers, I thank Chris Chappell, senior editor of Palgrave Macmillan USA, for his very patient guidance, understanding, and encouragement. I am grateful to Sarah Whalen, Mike Aperauch, assistant editors, for their hands-on supervision of the production process. I appreciate the thoughtful and carefully articulated comments of two anonymous reviewers that were useful to improve my draft.

This project was supported by several generous fellowships and grants, and I thank the following: OU's Research Degrees Committee for the three year research studentship; OU's Arts Faculty Research Committee for the extension of my studentship and for funding my short-term research trip to Japan in 2003; the Japan Foundation Endowment Committee, which generously funded my long-term research trip to Japan in 2004; CHF's Sidney M. Edelstein Fellowship, which also funded research trips to New York, Baltimore, and Washington, D.C.; the Postdoctoral Fellowship in the History of Modern Physical Sciences at MIT, which also included funding of research trips to New Haven, Connecticut; Ireland, and Japan; the Postdoctoral Fellowship in the History of Science and Technology in Modern East Asia at Harvard University, which also included funding of research trips to Oberlin, Ohio; and the IIAS affiliated fellowship.

I am grateful to the following libraries that supported my project: The Bodleian Japanese Library; the British Library; Cambridge University Library; CHF's Othmer Library of Chemical History; Harvard Yenching Library; Science Museum and Imperial College London Library; University Library and the East Asian Library, Leiden University; MIT Libraries; the National Diet Library; the Open University Library; the Library of the School of Oriental and African Studies, University of London; the University of Tokyo Faculty of Education Library; the University of Tokyo General Library; and the Widener Library, Harvard College Library. I especially thank the expert support of Noboru Koyama, curator of the Japanese collections at the

Cambridge University Library, and Kumiko Yamada McVey, librarian for the Japanese Collection at Harvard Yenching Library.

I would like to thank the archives listed under "Archival Sources" in the bibliography for their kind and competent support during my research. I am particularly grateful to the following: Yoshio Umezawa for allowing me to explore the archive of the Department of Chemistry, School of Science, University of Tokyo, as well as the records of the School of Science; the late Tetsuo Shiba for making the Sakurai Jōji Correspondence of the Chemical Society of Japan available to me; Hiroshi Motoyasu for the generous permission to explore freely the whole Sakurai Jōji Papers of the Ishikawa-ken Rekishi Hakubutsukan; and Tokuhei Tagai for giving me access to the notebook collection of Koto [the second o with macron] Bunjiro [o with macron] at the University Museum, University of Tokyo.

Portions of my book have previously appeared in various articles, and I am grateful to the editor and publisher of each journal for granting permission to use the materials here. Most of chapter 1 originally appeared as "Samurai Chemists, Charles Graham and Alexander William Williamson at University College London, 1863–1872," *Ambix* 56 (2009): 115–137; sections of chapter 2, chapter 4, and chapter 7 originally appeared in "Analysis, Fieldwork and Engineering: Accumulated Practices and the Formation of Applied Chemistry Teaching at Tokyo University, 1874–1900," *Historia Scientiarum* 18 (2008): 100–120; and a summary of two sections in chapter 6 appeared in "Cross-national Odyssey of a Chemist: Edward Divers at London, Galway and Tokyo," *History of Science* 50 (2012): 289–314, on 303–305.

My parents, Yasuyuki and Noriko Kikuchi, have constantly encouraged and supported my venture since I first talked with them about my wish to choose the history of science as my major some twenty years ago. My wife, Naoko Kikuchi, accompanied me on a long journey to this end. Her unwavering support is many-sided, but its most important aspect for me is that she has believed in the value of my work. I am grateful for her trust and love, and it is to her that I dedicate this book.

Abbreviations

Ann. Sci.	*Annals of Science*
Ayumi	Tokyo Daigaku Daigakuin Rigakukei Kenkyūka Rigakubu Kagaku Kyōshitsu Zasshikai, ed. 2007. *Tokyo Daigaku Rigakubu Kagaku Kyōshitsu no ayumi*
ACJ	*American Chemical Journal*
BAAS	British Association for the Advancement of Science
BJHS	*British Journal for the History of Science*
CN	*Chemical News*
Columbia College Catalogue	*Catalogue of the Officers and Students of Columbia College, for the Year*
Eng. Calendar Imp. Univ.	Imperial University of Japan (Teikoku Daigaku), *The Calendar for the Year*
Eng. Calendar Tokyo Univ. Law, Sci., and Lit.	Tokio Daigaku (University of Tokyo), *The Calendar of the Departments of Law, Science, and Literature*)
FCS	Fellow(ship) of the Chemical Society
Gs	*Gakugei shirin*
Hist. Sci.	*History of Science*
HSPS	*Historical Studies in the Physical (and Biological) Sciences*
ICET Calendar	Imperial College of Engineering, Tokei. *Calendar. Session*
JCSIUTJ	*The Journal of the College of Science, Imperial University (of Tokyo), Japan*
JCST	*Journal of the Chemical Society [London]: Transactions*
JHU	Johns Hopkins University, Baltimore, Maryland, USA
JHUC	*Johns Hopkins University Circulars*
Jpn. Calendar (Tokyo) Imp. Univ.	*(Tokyo) Teikoku Daigaku ichiran*

Jpn. Calendar Kyoto Imp. Univ.	*Kyoto Teikoku Daigaku ichiran*
Jpn. Calendar Tohoku Imp. Univ.	*Tōhoku Teikoku Daigaku ichiran*
Jpn. Calendar Tohoku Imp. Univ. Coll. Sci.	*The Tohoku Imperial University College of Science Calendar. Tōhoku Teikoku Daigaku Rika Daigaku ichiran*
Jpn. Calendar Tokyo Imp. Univ.	*Tokyo Teikoku Daigaku ichiran*
Jpn. Calendar Tokyo Univ. Law, Sci., and Lit.	*Tokyo Daigaku Hō- Ri- Bun-Sangakubu Ichiran*
JSA	*Journal of the Society of Arts*
JSHS	*Japanese Studies in the History of Science*
KCL	King's College, London, UK.
KDh	Kyoto Daigaku hyakunenshi henshū iinkai, ed. 1997–2001. *Kyoto Daigaku hyakunenshi*
Kk	*Kagakusi kenkyū*
Knr	*Kagaku no ryōiki*
Kōbunroku	Kokuritsu Kōbunshokan, Tokyo, Japan, *Kōbunroku* [Central government documents in early Meiji Japan].
Monbushō ōfuku	Tokyo Daigakushi Shiryōshitsu, Tokyo, Japan, *Monbushō ōfuku* [Tokyo University Correspondence with the Ministry of Education].
Nakazawa Iwata kinenchō	Nakazawa Iwata hakushi kiju shukuga kinenkai, ed. 1935. *Nakazawa Iwata hakushi kiju shukuga kinenchō.*
NDSB	Koertge, Noretta, et al., eds. 2007. *New Dictionary of Scientific Biography*
Nkgt	*Nihon kagaku gijutsushi taikei*
Owens College Calendar	The Owens College, Manchester, *Calendar for the Session*
RCC	Royal College of Chemistry, London, UK.
RSM	Royal School of Mines, London, UK.
Sakurai Jōji Correspondence	Nihon Kagakukai, Correspondence from abroad of Sakurai Jōji (ex Sakurai Sueo)

Sakurai Jōji Papers	Ishikawa-ken Rekishi Hakubutsukan, *Sakurai Jōji Kankei Shoshiryō*
Shiryō oyatoi gaikokujin	UNESCO Higashi Ajia Bunka Kenkyū Sentā, ed. 1975. *Shiryō oyatoi gaikokujin*
Takamatsu Toyokichi den	Takamatsu Hakushi Shukuga Denki Kankōkai, ed. 1932. *Kōgaku Hakushi Takamatsu Toyokichi den*
TASJ	*Transactions of the Asiatic Society of Japan*
TDh	Tokyo Daigaku Hyakunenshi Hensan Iinkai, ed. 1984–1987. *Tokyo Daigaku hyakunenshi*
TDn	Tokyo Daigakushi shiryō kenkyūkai, ed. 1993–1994. *Tokyo Daigaku nenpō*
Tgz	*Tōyō gakugei zasshi*
Tkk	*Tokyo kagaku kaishi*
Tsuiokuroku	*Ikeda Kikunae hakushi tsuiokuroku* (1956).
TTDg	Tokyo Teikoku Daigaku. 1932. *Tokyo Teikoku Daigaku gojūnenshi*
UCL	University College, London, UK
UCL Calendar	University College, London, *Calendar for the Session*

Note on Conventions

Japanese names in what follows are written in the order used in Japan, with family names preceding given names. The way of referring to historical figures in this study can be confusing because this book covers both the pre- and post-Meiji-Restoration periods for which different conventions are normally used. I have adopted the standard *post*-Meiji-Restoration custom of referring to people by their family names throughout. In transliterating Japanese words I have adopted the modified Hepburn-system (*hebon shiki rōmaji*) as is shown in Watanabe Toshirō, Edmund R. Skrzypckak, and Paul Snowden, eds., *Kenkyusha's New Japanese-English Dictionary*, 5th edition (Tokyo: Kenkyūsha, 2003), p. xiv. I have shown transliterated Japanese words in italics with two exceptions: (1) proper nouns are written without italicizing, such as Itō Hirobumi and the Tokyo Kaisei Gakkō; and (2) words well-known to English readers, such as Tokyo, Osaka, Kyoto, sake, and samurai, are written with neither italicizing nor macrons. I basically follow Janet Hunter, ed., *Concise Dictionary of Modern Japanese History* (Berkeley, Los Angeles, and London: University of California Press, 1984) in rendering technical terms of Japanese history into English.

Introduction

The object of this book is *not* to give the *whole* picture of the transmission of Western chemistry to modern Japan. Instead, this book is concerned with how scientific practices in different parts of the world are connected with each other. It is part of my conscious move away from the grand narrative of "Japan meets the West" or "the East meets the West," which has for so long affected the historiography of modern science in East Asia.[1] The strategy I am adopting is to stress the highly localized and temporized nature of the science by focusing on chemistry in Britain and the United States in the second half of the nineteenth century. The other more important aspect of this book is to look at human encounters in physical spaces. Scholars have identified various kinds of human agents as an essential medium of global circulation, communication, and appropriation of knowledge, which has inspired my approach.[2] As will become clear, however, I take spaces unusually seriously as an independent agency because they fundamentally affect the nature of human encounters. This approach is summed up in the subtitle of this book, "the lab as contact zone." By applying the concept of the contact zone this book sheds new light on the relevance of spaces and localities to scientific practice.[3] The spaces are not only sites of the production, transmission, and legitimization of knowledge and materials—an aspect I never underestimate—but they are also sites of human encounter and the resulting intricate social interactions that in turn affect the shape of scientific practice.

There were several possibilities of such human encounter for Japan during the period under discussion. The opening of treaty ports in the 1850s and 1860s meant increased contact between foreign and domestic traders, their clients, and other personnel there.[4] A long-standing ban on foreign travel was lifted in 1866, after which any Japanese could apply for permission to stay "for study and trade" in countries that had concluded treaties with Japan, among them major European countries and the United States.[5] Early on, Japanese overseas travelers were predominantly students.[6] Likewise, probably the easiest way for ordinary Japanese to get acquainted with foreigners would have been to become a college student, though even this option was out of reach for most of the populace, especially for women. In the nascent nation-state trying to build a Western-style higher education system as part of its industrialization policy, the new Japanese government established by the Meiji Restoration in 1868 hired a large number of foreign teachers and sent the most successful students abroad for overseas study. This is why I focus on scientific pedagogy—on how young scientists are made—in this book.

The other important reason for this focus is the long-term effect of scientific pedagogy on the development of scientific practice. Historians, philosophers, and sociologists of science have begun to look critically at scientific pedagogy as one of the central issues in science and technology studies.[7] Recent work has highlighted three overarching questions: (1) how and to what extent does scientific pedagogy shape scientific research? (2) How different from place to place is a "pedagogical regime," a social fabric that affects and controls what is taught, how a scientist is trained, and what role models and rules of conduct are honored? (3) How does a pedagogical regime interact with society at large? In light of today's global and transnational societies, it is necessary, even pressing, to add to this research agenda a fourth dimension: cross-national exchange of ideas, people, and materials for the construction of a pedagogical regime.[8] Considering complex national and regional cultural differences, we cannot assume that there is one single answer to the question of how best to train scientists and engineers. However, we can learn from historical case studies how educators in the past managed similar kinds of problems and reached some solutions.

Arguably the first laboratory-based science, chemistry has had a distinct history since antiquity as the hybrid of craft and philosophical activities where material production and the production of knowledge of matter are inseparably entangled.[9] Until the end of the eighteenth century, the material culture of the chemical laboratory was virtually identical with that of workshops and factories. It was in the first half of the nineteenth century that these two spheres started to separate somewhat, albeit far from completely, with the spread of the Lavoisian theory of chemical elements and the Daltonian/Berzelian atomic theory across Europe and North America (together with their linguistic and graphic representations) and the miniaturization and standardization of chemical analysis, which logically assumed some kind of elemental and/or atomic theories.[10]

It is for this historical reason that chemistry was widely considered the most practical and utilitarian of all scientific subjects throughout the nineteenth century and played a big role in the contemporary debates regarding "applied science" and "technical education" on both sides of the Atlantic.[11] In other words, chemistry was a distinct and exemplary nineteenth-century discipline among others, such as natural philosophy (physics) and biology, on which chemistry had a great impact.[12] Therefore, it cannot be considered a mere coincidence that chemistry was one of the first academic subjects institutionalized in Japanese higher education (together with law and engineering subjects), as I discuss later on in this book. The Japanese case instead underlines the truly international character of the debates about applied science and technical education and merits the serious attention of historians of science and technology.

Japan in the Meiji period (1868–1912) desperately needed foreign support to fill the perceived need for a Western-style system of higher education, something that was essential for survival in an increasingly Euro- and American-centric world order after the First Opium War of 1839–1842. Japan's political leaders and educators first flocked to Britain as a strong industrial sea power and to the United States as the country that triggered the whole process of Japan's opening of diplomatic relations with world powers.[13] Despite the long-lasting image of German science as a model for Japanese science from the Meiji period onward, British and American teachers were dominant in Japanese higher education between the 1860s and 1880s, and many Japanese overseas students went to British and American universities and colleges to finish their training during this period.[14] Increase of German presence in Japanese higher education (and in politics and administration) came later, from the 1880s onward. Before then, the German presence was restricted to the medical sciences, including medical and pharmaceutical chemistry, because Dutch-style medical education, the strongest specialty of Dutch learning in Japan during the Tokugawa period, was Germanized in the early 1870s.[15] As a result, Meiji Japan became a kaleidoscope of European and North American (as well as Japanese) styles in many aspects of institutional as well as material culture.

Chemistry was at the center of this educational development. Indeed, British chemists collaborated with American chemists in setting up chemical education at Japanese universities and colleges, and Japanese chemistry students in Britain and the United States became the first Japanese chemistry professors. These facts are known among historians of science in Japan on both sides of the Pacific and on both sides of Eurasia. But the *process*—how British and American connections in Japanese chemical education emerged, were sustained, and later diminished—and its *consequences* for Japanese science and technology has not been examined yet.

This book focuses on these important developments and argues that the Anglo-Japanese and American-Japanese connections in chemistry had a major impact on the institutionalization of Japanese scientific and technological higher education from the late nineteenth century onward; they helped define the structure of the Japanese pedagogical and research system that lasted well into the post-World War II period of massive technological development, when Japan became one of the biggest providers of chemists and producers of chemical publications in the world next to the United States, Soviet Union, and Western Europe.[16]

As for the first theme of the *process*, I argue that among the earliest Japanese chemistry students in Britain and the United States in the 1860s were influential members of the Meiji government such as Itō Hirobumi and Mori Arinori. Their views on science, technology, and education were shaped through interaction with their British and American teachers, and these students became instrumental in implementing the government's industrialization policies, a major part of which was higher education. These students developed Anglo-American scholarly relations in chemistry throughout the 1870s. These involvements of politicians and administrators gradually gave way to the choices of chemists themselves in the late 1870s and 1880s, some of whom favored Germany while others kept close ties with Britain and/or the United States. How this happened provides the basic story line of this book.

I interpret the second theme of the *consequences* of Anglo-Japanese and American-Japanese relations in chemistry for Japanese science and technology as the issue of cross-cultural transfers of knowledge and models within what Mary-Louise Pratt called contact zones. She defined it as "a space in which peoples geographically and historically separated come into contact with each other," and in which "transculturation"—the phenomenon of a "union of cultures" or merging and converging cultures—occurs.[17] According to Pratt,

> A "contact" perspective emphasizes how subjects are constituted in and by their relations to each other. It treats the relations among colonizers and colonized, or travellers and "travelees," not in terms of separateness or apartheid, but in terms of copresence, interaction, interlocking understandings and practices, often within radically asymmetrical relations of power.[18]

As is clear from this quotation, Pratt uses the concept of contact zones against the backdrop of colonial encounters. She carefully disentangles the dictionary definition of a contact zone from formal colonialism, but this recognition of sheer power inequality and the possibility of interaction in spite of it distinguishes "contact zones" from other similarly anthropology-inspired concepts such as Peter Galison's "trading zones," the space where people with different professional cultures, such as physicists, engineers, and instrument makers mingle, negotiate, and exchanges ideas, instruments, and data.[19]

This aspect of contact zones is relevant to my case in two ways. First, though Japan in the late Tokugawa and Meiji periods narrowly escaped formal colonialism, much of the actions of those in power were dictated by the painful recognition of power inequality (both military and economic) vis-à-vis Euro-American world powers, which is something akin to what Michael Herzfeld called "crypto-colonialism," a hidden form of colonialism by means of cultural, economic, and epistemological hegemony.[20] Second, the pedagogical situations I am about to analyze—in the classroom, teaching laboratories, etc.—are inherently marked by unequal power relationships of some sort (overt, disguised, benign, or subverted) to varied extents between teachers and students.[21] Both kinds of dynamics are nicely captured by contact zones.

In examining the mechanism of transculturation of educational models by analyzing interactions between British and American chemistry teachers and their Japanese students in the Meiji period, this book focuses on several pedagogical spaces qua contact zones to highlight the importance of place, mutual give-and-take, and conflict-solving in cross-cultural interaction. Here we are looking at two kinds of interactions in the same contact zone: students' appropriation of teachers' professional culture (the very essence of scientific pedagogy), and cross-national interaction between Japanese students and British and American teachers. This approach thus emphasizes personal contacts as the medium of transmitting chemical expertise *and* educational models, and it also brings the pedagogical regimes prevailing in British, American, and Japanese chemistry education together into the spotlight.

In light of the complexity of scientific pedagogy and the importance of laboratory training in chemistry, it is simply not enough to stay with classical topics, such as curriculum design, lecture notebooks, and textbooks as components of an educational model. Contact zones inform my approach by drawing attention to the manifold and inseparable relationships between pedagogy, research, and the material construction of pedagogical spaces that had a far-reaching impact on connecting—and separating—people, cultures, disciplines, subdisciplines, and research units. The crucial point here is the duality of a contact zone. First, as an actual physical space defined by a material building and furnished with a variety of equipment, such as glassware and gas, and second as a means for constructing social relations.

This duality, together with the fact that the contact zone as a concept has remained a heuristic device and not been used as an analytical tool to interpret actual spaces, makes it necessary to marry the concept with other conceptual and empirical tools to bring to bear its full potential on elucidating how a space and place affect scientific practice and pedagogy and vice versa.[22] There are two points of reference in my approach to this question that correspond to this duality. The first is the early Meiji educationists' holistic approach to material culture. That is, their attitudes toward scientific pedagogy and architecture, for example, did not exist in isolation but was part of their overall attitude toward British or North American material culture. This is indeed congruent with Jules David Prown's definition of material culture as "the manifestations of culture through material production."[23] However, we cannot assume automatically that historical actors thought or felt this way; our understanding has to be empirically grounded. That is part of what I do in chapters 2, 4, and 5.

The second point of reference is the Foucauldian analysis of architecture that looks at supervision, surveillance, and other means of social control as key tools for analyzing architecture and the spaces it helps create. This perspective was inspired by the panopticon, the prison with a tall watch tower surrounded by cells, conceived by English utilitarian philosopher Jeremy Bentham.[24] Recent architectural historians and urban designers, however, have started to depart from Michel Foucault's pessimism and have begun to treat supervision and surveillance more as the means of allowing effective social interaction among residents in urban spaces.[25] Likewise, I am concerned with the question of who stood at the center of the network of human relations between teachers and students that was created by pedagogical spaces as contact zones.

Accordingly, in what follows I often use the term "assistant-centered structure" vis-à-vis a "professor-centered structure" to show my answer to this question in a particular space. The former refers to a pedagogical space, such as a teaching laboratory, where assistants, not professors, have daily contact with students. In contrast, in the latter, professors impart knowledge to students, such as in a lecture hall, and assistants play the role of true "subordinates" helping professors' demonstrations. This does not preclude the possibility of one assistant playing both roles, but it is important to distinguish two very different pedagogical spaces and situations.

This distinction also enables us to highlight the importance of assistants and advanced students within the "assistant-centered structure" as "cultural mediators" between foreign-trained professors and Japanese students, which has an obvious parallel with what Simon Schaffer, Lissa Roberts, Kapil Raj, and James Delbourgo called "go-betweens."[26] This structure largely determined what kind of people Japanese chemistry students would be most likely to interact and work with. In this way, the contact zones functioned both as a medium and a model, and that is why they are important to my story.

Thus far I have put aside another important aspect of cross-cultural interaction: communication within contact zones. It is to unravel the haphazard and at times destructive but occasionally also productive aspects of cross-cultural communication that I introduce the fourth interpretative tool, "Translatability of Culture." It was born out of interdisciplinary discussions among researchers in philosophy, fine arts, music, and literature as well as in science and technology studies.[27] The basic idea here is to look at the process of unavoidable semantic changes caused by two-way processes of translation of cultural elements and the associated cognitive problem of "incommensurability" of different cultures. This approach emphasizes the dual meaning of the word "translation": (1) to change speech or writing into another language, and (2) to move something from one context to another, thus making it possible to talk about "translation" of nonverbal cultural elements, such as painting and music. This double meaning of translation is particularly relevant to my case of the transculturation of British or North American models of scientific pedagogy in Japan.

Even if one succeeds in finding or coining more or less appropriate Japanese words to express an idea of British or American origins about scientific pedagogy, the meaning may be unintelligible, destroyed, or totally changed in the Japanese context, just as might be the case for foreign art objects and musical compositions transplanted into the Japanese context or vice versa. That is where paraphrasing or explanation with languages familiar to Japanese audiences begins. As I argue in the following chapters, particularly in chapter 4, that was exactly what Japanese scientists had to do to varying degrees; skilful and minute adjustments were in some cases not only unavoidable but the key to succeeding in the transculturation of British or American models of scientific pedagogy in Japan.

The chapters that follow are laid out roughly in chronological order, but each has a particular thematic focus. Chapter 1 is about Anglo-American contact zones where Japanese students from Chōshu and Satsuma domains (*han*), without any experience in Western-style scientific pedagogy, interacted with British and American teachers and formed their views on science, technology, and education in the 1860s. Their views mattered because these two domains, semi-autonomous provinces owned and ruled by feudal lords, soon toppled the Tokugawa Shogunate in 1868 and formed the core of the new Meiji government, providing influential politicians and educators. Those students' views are good examples of Japanese reactions to the contemporary debate in Britain on technical education, because two of their teachers, Alexander William Williamson and Charles Graham, were active participants in the debate. I examine a variety of contact zones and elucidate how such spaces affected Chōshū and Satsuma students differently. I also introduce the term "assistant-centered structure" of laboratory contact zones at UCL, which plays an important role later in this book. I finish this chapter with the other major difference between Chōshū and Satsuma students, i.e., Satsuma students' subsequent move to the United States and study at Rutgers College in New Brunswick, New Jersey.

In chapter 2 we move to contact zones in Japan. The chapter scrutinizes one of the earliest cases of laboratory-based chemical education in early Meiji period, the Department of Chemistry led by British chemist Robert William Atkinson at the Tokyo Kaisei Gakkō and Tokyo University in the 1870s. Though located in Tokyo, this school had a multinational faculty. Its classrooms and laboratories effectively became contact zones whose physical spaces exhibited a hybrid material culture. I elucidate how Atkinson used the legacy of his American predecessor William Elliot Griffis, his own experience in London, and Japanese cultural elements (Japanese indigenous manufactures of sake and soy sauce, for example) to construct his style of chemical education as an example of transculturation. This chapter also highlights the crucial roles of cultural mediators, such as school director Hatakeyama Yoshinari (a Satsuma student at UCL) and the assistant to Atkinson, Masaki Taizō (another Japanese student at UCL, from Chōshu).

Chapter 3 is about Japanese students who experienced both Japanese and Anglo-American contact zones. There were four chemistry students from the same Tokyo Kaisei Gakkō and Tokyo University who later moved to different places in Britain and the United States for overseas study: Matsui Naokichi (Columbia College School of Mines), Sakurai Jōji (UCL), Takamatsu Toyokichi (Owens College Manchester), and Kuhara Mitsuru (JHU). They have been chosen on the basis of their subsequent roles in the establishment of scientific and technological education in chemistry in Japan. The central question is how those four chemists developed totally different visions of chemical education despite their common origin in Tokyo. The chapter also emphasizes Masaki's continued role as cultural mediator between Japanese students and British teachers in his capacity as superintendent of overseas students in London.

Chapter 4 follows the career paths of the same protagonists we met in chapter 3 after they returned to Japan to take up positions at their alma mater. I tell the story against the backdrop of arguably the most important restructuring process in the history of Japanese higher education, namely, the merger of Tokyo University and another prominent engineering school, the Imperial College of Engineering, Tokyo, into the Imperial University in 1886. In that chapter I focus on several "translation" strategies (ranging from "literal" translation to careful cultural adjustment) with which the former students used their experiences in Britain and the United States to negotiate for and construct their own pedagogical spaces between 1880 and 1886. The languages they used such as distinctly Japanese "manufacturing chemistry" (*seizō kagaku*) and "pure and right chemistry" (*junsei kagaku*) show the variety and subtlety with which the Euro-American rhetoric of "applied science" was negotiated in a different locality. I conclude this chapter with a section on the organizational structure of the Imperial University and describe how its high-ranking administrators looked at the place and meaning of chemistry within the Japanese system of higher education and society more generally and how their views affected the activities of chemistry professors.

In the next three chapters I examine the two major products of the restructuring process examined in chapter 4. Chapter 5 addresses the question of the kind of pedagogical space Sakurai, with support of the British chemist Edward Divers, constructed for the Department of Chemistry at the Imperial University in Tokyo in the late 1880s and 1890s with the Foucauldian analysis of architecture as a starting point. The main question here concerns the way Sakurai juxtaposed classrooms, laboratories, and office spaces and thus structured contact zones between teachers and students. What stands out in Sakurai's and Divers's pedagogical space is the strong influence of the British "assistant-centered structure," where the junior faculty had an important role as experimental trainers and as cultural mediators, in spite of the German origin of the "departmental space." That chapter also reveals a distinctly cross-cultural element: the *Zasshi-kai* (journal meeting), an informal discussion and socializing event held weekly at the library that combined Anglo-American style reading seminars with Japanese leisure culture.

Any study of a pedagogical space would not be complete without looking at what actually occurred in it. Chapter 6 looks at how differently these two senior professors at Tokyo's Department of Chemistry made use of the same space analyzed in chapter 5 to deliver lectures, to supervise advanced students' works, to do their own research, and to connect to students in other ways. The chapter also examines how the students, most notably Ikeda Kikunae (physical chemistry), Haga Tamemasa (inorganic chemistry), and Majima Toshiyuki (organic chemistry), responded to the teaching both as followers and as dissidents.

Chapter 7 is about another product of the restructuring process, the Department of Applied Chemistry of the College of Engineering. Again, it was an outcome of cross-cultural interaction between Britain, Germany, and Japan. Takamatsu and his German-trained colleague Nakazawa Iwata set up a teaching program of applied chemistry for the Imperial University and technical colleges in the late 1880s and 1890s. They had to meet the special challenge of reaching out to the manufacturing sector that lay outside the hierarchy of the national educational system. As a result, the Imperial University developed a pair of completely separate departments for pure and applied chemistry with different teaching philosophies and social networks.

The brief epilogue summarizes theoretical innovations in this book and draws conclusions from it. It also looks at Kuhara, Majima, and other Japanese chemists trained at Tokyo Imperial University who held positions at the next generation of imperial universities, such as Kyoto and Tōhoku (Sendai), at the turn of the century to show a wider nationwide effect of this story in the twentieth century. Focusing on the role of intergenerational and intragenerational conflicts, the epilogue traces how these chemists reused the legacy of their teachers and yet partly deviated from them in the construction of pedagogical spaces and other aspects of teaching, resulting in the "fiefdom" structure that was later misleadingly called the *kōza* (chair) system and in a blurred but still existent distinction between pure and applied chemistry.

One last word to clarify what I mean by "the highly localized and temporized nature of the science." One example of such temporized and localized culture is the teaching chemical laboratory, which is so ubiquitous all over the world today. There was not even a beginning of it to be found in any part of Europe and America in the eighteenth century. In addition, it is misleading, if not completely wrong, to assert that this institution simply spread from Justus Liebig's chemical laboratory in the Hessian town of Giessen to the whole of Europe and North America in the mid-nineteenth century.[28] The cross-national "genealogy" of teaching chemical laboratories is much more complex. Like a common theme with variations, this development played out in a variety of local, national, and cross-national contexts.[29] It is my hope that readers get a renewed sense of such complexities through the lense of *Anglo-American Connections in Japanese Chemistry*.

Chapter 1

Japanese Chemistry Students in Britain and the United States in the 1860s

Two pioneering groups of young Japanese studied chemistry and other scientific subjects at UCL in the 1860s. They came to London in 1863 and 1865 from the Chōshū and Satsuma domains (*han*) in Western and Southwestern Japan, respectively. Some of the Satsuma students went even further across the Atlantic and ended up studying at Rutgers College in New Brunswick, New Jersey. It was a treacherous business in many ways. In addition to being on the sea for such a long time, which was much more dangerous in the 1860s than it is today, it was illegal under the laws of the Tokugawa Shogunate (*Bakufu*). By sending retainers abroad, these two domains defied the long-standing ban of overseas travel imposed by the shogunate and eventually joined force with other anti-Tokugawa domains to overturn that shogunate's regime and usher in the Meiji Restoration in 1868.[1]

At first glance, this event might seem to be a false start of our story about scientific pedagogy of chemistry in Meiji Japan. In fact, it is a perfect example of how nonexperts have a tremendous impact on scientific pedagogy as politicians and administrators. While former retainers of the Chōshū and Satsuma domains formed the core of the new Japanese government under Emperor Meiji (reigned 1867–1912), several members of these groups later held senior government and academic positions and were consequently involved in the employment of British and American chemists and the overseas study of Japanese chemists in Britain and the United States. Therefore, in addition to their central roles in Japanese history, their views on science, technology, and education are important for understanding how Anglo-Japanese and American-Japanese scholarly relations in chemistry developed, and for understanding the general process of institutionalization of scientific and technological education in Meiji Japan.[2]

Unlike later students in the Meiji period, the first students from the Chōshū and Satsuma domains had to make sense of what they were doing in Britain and the United States on their own without any previous experience of Western-style schooling and with little firsthand information on scientific pedagogy in Western institutions. Thus, their experience was a quintessential example of cross-cultural interaction in contact zones. As I discussed in the introduction, I am concerned with the question of who stood at the center of the structure of contact zones or the network of human relations between UCL chemistry teachers and Japanese and other students. The Japanese connections of Alexander William Williamson, the professor of chemistry and practical chemistry, are relatively well documented.[3] This includes the assumption that because he was at the top of the hierarchy of UCL chemical education, he was at the center of the contact zones between teachers and students as well. This "professor-centered" structure applies to a lecture hall, for example, where a lecturer was a communicator and students tended to be passive recipients. However, both Chōshū and Satsuma students generally first opted to attend laboratory courses, where assistants and advanced students played distinctive roles as intermediaries between professors and junior students.[4] It seems that there was an "assistant-centered" structure of contact zones between Japanese students and UCL chemists.

In light of the above consideration, I aim to answer the following three questions in this chapter: (1) how did the encounter between Chōshū and Satsuma students and their British chemists affect their views on science, technology, and education on both Japanese and British sides? (2) How did the structures of contact zones between them affect Chōshū's and Satsuma's experiences and views? (3) How did Satsuma students' subsequent move to the United Stated and study at Rutgers College affect these differences? To gauge the impact of their encounter, one has to profile Chōshū and Satsuma students and their British and American teachers before the encounter, which I do as a preamble to the analysis of their encounter, contact zones, interactions, and aftermath. I also address the question of what brought them to UCL as part of this preamble.

Much of the discussion between Japanese students and their British and American teachers covered in this and the following chapters was centered on how to strike a proper balance between the scientific and technological components of technical education, that is, how to educate and train a good technologist. We are here looking at a question of global importance throughout the nineteenth century and beyond in Britain and the United States as well as in Japan.[5] I argue in this chapter that *both* Japanese students and their British and American teachers learned from each other because neither Britons nor Americans could give the Japanese students a definite answer. This situation, in addition to the structures of contact zones encouraging mutual dialogue, made meaningful interaction between them possible.

Students' Motives and Aspiration for Studying in Britain

The motives of the Chōshū and Satsuma governments in sending students to Britain and the aspiration of the students who were sent should be understood in the light of two contexts in which their educational plans were made and executed: the militarization of Western learning and the *sonnō jōi* ("revere Emperor, expel barbarians") movement that spread among samurai (the ruling warrior class) in the Bakumatsu period (i.e., the late Tokugawa period, ca. 1850s to 1868).

Dutch learning and Western learning developed in Japan from the eighteenth century onward mainly through activities of medical doctors and astronomers.[6] From 1842 on, however, when news of the defeat of Qing China in the First Opium War reached Japan, a growing sense of international crisis arose among samurai and led to a strong interest in Western learning, which as a result underwent what historians of Western learning in Japan called "militarization" (*gunji kagakuka*). In the Chōshū domain, this trend toward Western military learning was intertwined with the xenophobic *jōi* movement, which was prompted by the Tokugawa Shogunate's action in 1858 to conclude international treaties with the United States, the Netherlands, Russia, France, and Britain without approval from the Emperor.[7] The Chōshū domain had been the very center of *jōi* movements, which had been adopted as a policy of the domain. These circumstances inevitably influenced the feeling of the military-minded Chōshū students who would go to Britain in 1863. For example, when Takasugi Shinsaku, the leading military reformer of the Chōshū domain, led an arson attack on the newly built British embassy building in Shinagawa near Edo (Tokyo) in December 1862, he was joined by future UCL Chōshū students, Inoue Kaoru, Itō Hirobumi, and Yamao Yōzō.

Therefore, one important motive of the Chōshū domain government in sending students overseas was to have them learn naval techniques to "suppress barbarians with the arts of barbarians" (*i no jutsu o motte i o seisu*). Sending young samurai to Britain, the foremost sea power throughout the nineteenth century, was a sensible choice for the Chōshū leadership. Senior government officials of the domain secured funding for overseas students by diverting money allocated for purchasing Western-style guns, and likened sending the students to the purchase of "human machines" (*hito no kikai*) and "live machines" (*ikeru kikai*). This epitomized the officials' keen interest in Western military technology and their image of both Western civilization and samurai youth as commodities.[8]

The Satsuma students in Britain shared the Chōshū students' interest in naval and military techniques, and some also had strong *jōi* sentiments. Students' interests were reflected in their proposed (but largely unrealized) subjects of study in Britain, which included naval and military studies, fortification, gunnery, shipbuilding, and naval surveying.[9] However, Satsuma's domain government was firmly against the *jōi* movement. The Satsuma slogan "enriching the nation, strengthening the army (*fukoku kyōhei*)," was itself commonly used in various domains, including Chōshū, and in the Tokugawa Shogunate. It gained general currency during the Meiji period. Yet Satsuma's version emphasized "enriching the nation," which the domain government thought was possible only by fostering Western-style manufacturing technology and encouraging trade with foreign countries. This argument gathered momentum especially after the bombardment of Kagoshima (the capital of the Satsuma domain) by the British Royal Navy in 1864. The intention of the Satsuma overseas study plan was thus more broadly oriented toward learning manufacturing technologies from an industrialized nation in addition to military technology. These twin goals reflected the Westernizing "enriching the nation" policy of the plan's mastermind, Godai Tomoatsu.[10]

Connecting Japanese Students to UCL and Williamson

The two domain governments used their connections with British merchants in treaty ports when arranging overseas study for their retainers.[11] The student group from Chōshū went to Britain in 1863 with the assistance of Jardine, Matheson, and Company, a Far East trading company from which the Chōshū domain purchased ammunition. Hugh M. Matheson, the company's London agent, welcomed the students and made arrangements for them. Thomas Blake Glover, a Scottish merchant who later had dealings with the Satsuma domain, was the main intermediary of Satsuma overseas study in 1865.[12] Ryle Holme, Glover's associate, accompanied the Satsuma group, and Glover's brother, Jim, made arrangements for them in London. On the academic side, the principal organizer for both groups was Alexander William Williamson, UCL's professor of chemistry and practical chemistry who took responsibility for the students' private and social lives as well as their academic lives in London by being their teacher, advisor, and landlord.

There is no direct evidence of any preexisting links between the British merchants and Williamson. Matheson recalled in his autobiography: "I was extremely fortunate in inducing Dr Williamson, Professor of Chemistry in University College, afterwards President of the British Association, to receive them into his house," but he did not explain how he became acquainted with Williamson or why he chose Williamson for the task.[13] However, Williamson's obituary, written by his former student and physics colleague George Carey Foster, speaks of "the recommendation of Mr (now Sir Augustus) Prevost," an auditor of University College London, who influenced placing "five young Japanese noblemen ... under Williamson's care by Mr Matheson, of the firm of Jardine, Matheson and Co."[14] It would seem that Matheson turned initially to UCL, rather than to Williamson personally.[15]

The facts that UCL was an educational institution delivering scientific and practical instruction and that it was secular in character were not sufficient to account for Matheson's choice. For example, the RCC (which was by then RSM's Chemistry Department) and Owens College Manchester met the same criteria. UCL, however, had uniquely established Far Eastern and Asian connections. For example, the UCL Council occasionally received letters from Lord John Russell, foreign secretary in the 1860s, asking UCL to recommend students who might become student-interpreters, an entry-level position for diplomats in Far-Eastern countries such as China and Japan.[16] The example of Ernest Mason Satow, an ex-student of UCL who pursued a career as a diplomat in Japan and China, is well known.[17] UCL, as an institution, had fostered Far Eastern connections that were recognized by the British Foreign Office.

Financial circumstances at UCL promoted acceptance of Japanese students as well. A severe financial situation induced the college's council to establish, in March 1865, an internal committee to discuss possible measures to "increase the income of the College by augmenting the number of Students and Pupils."[18] Eighteen months earlier, the college had received a £3,000 donation from Cama and Company, an Indian trading company that praised the college's "principle of imparting the best literary and scientific instruction to all Students without distinction of religious opinions."[19] However negative the financial circumstances may sound, they created a favorable condition for potential international students, including the Chōshū and Satsuma, to receive education at UCL.

Williamson's cosmopolitan background probably contributed to his enthusiasm for hosting Japanese students. His father, also an Alexander, joined the East India Company in 1810 as a clerk in the Tea and Drugs Warehouse and later worked as assistant to the Searcher of Records under the Examiner of Indian Correspondence between 1819 and 1834.[20] Williamson senior was a friend of his superior at the East India Company, the Examiner of Indian Correspondence, James Mill, and his son John Stuart Mill.[21] Alexander Williamson junior was familiar with utilitarian philosophy through these family connections. I interpret this as a key to understanding Williamson's interest in Asia and his cosmopolitan outlook exemplified in his inaugural lecture as professor of practical chemistry at UCL in 1849.

Williamson's lecture, entitled "Development of Difference the Basis of Unity," is at first sight an ill-organized address based on impressionistic observations of several European nations. It was not well-received by its audience.[22] Yet, with simple logic he developed his arguments, namely, that the greater the cultural difference between individuals or nations, the more advanced their civilizations.[23] Williamson praised the diversity of different cultures in the world as a sign of the advancement of world civilization and stressed the need for "union of difference" by understanding and admiring such cultures that, Williamson asserted, was the proper mission of UCL.

Williamson's cosmopolitanism would not treat each country or region equally, however. Using the same logic, he also compared religious tolerance and pluralism in Britain favorably with the Catholicism of, say, France or Spain, and Europeans favorably with "the native of New Holland," i.e., Australian Aborigines.[24] In this light Williamson's perceived mission of UCL can be construed as a colonial strategy to augment a sense of British and European cultural superiority to other parts of the world.

Williamson and Graham as Chemistry Teachers: The Liberal Science Model

The underlying ideas in the teaching of Williamson and his laboratory assistant Charles Graham, who will play an important role in this story, and the nature of the education that Chōshū and Satsuma students were to receive at UCL are best understood in terms of what Robert Bud and Gerrylynn Roberts called the "liberal science model."[25] According to this model inspired both by the English tradition of liberal education at Oxford and Cambridge and laboratory-based chemistry teaching practiced by German chemist Justus Liebig at Giessen, the principal aim of chemistry teaching at universities and colleges was not vocational training in a particular industry but liberal education for a wider audience. It was designed to discipline students' minds and hands through systematic learning of the theoretical principles of pure chemistry and other scientific subjects such as physics and geology and through practical training in generic chemical analysis; this, it was argued, would provide a sound basis for subsequent employment in a wide variety of chemistry-related fields, such as pharmacy, medicine, agriculture, metallurgy, manufacturing, and teaching.

Williamson's chemistry teaching at UCL, as shown in his addresses, syllab uses, and examination papers, reveals these key elements of the liberal science model.[26] His philosophy of proper university science education is best expressed in *A Plea for Pure Science*, his inaugural address in 1870 as the first dean of UCL's Faculty of Science. For him science was a mode of enlarging the mental faculties, its object being "to systematize our knowledge" by knowledge of general principles and laws and to apply such knowledge "to the explanation of the observations which had been made, and to the anticipation of others."[27] Williamson believed that chemical technology, which he regarded as "the application of scientific principles to the complex processes which occur in manufacturing arts," should and could only be taught outside colleges in real workplaces.[28]

Charles Graham in his student days at UCL was strongly influenced by of Williamson and his liberal science model. Born in Berwick-upon-Tweed in 1836, Charles Graham first attended Nesbit's Chemical and Agricultural College in Kennington, South London to study analytical chemistry and agricultural chemistry in particular before entering UCL in 1863.[29] He took botany, zoology, geology, mathematics, chemistry, and analytical chemistry there between 1863 and 1865.[30] He was awarded a bachelor of science degree with honors in chemistry and natural philosophy (in his first examination) and in geology and paleontology (in his final examination) from the University of London in 1864.[31] Graham concentrated on studying scientific subjects at UCL in accordance with Williamson's liberal science model. Graham was appointed assistant to Williamson around this time, and he was awarded a doctor of science degree from the same institution in 1866 while tutoring Japanese students.

Japanese Students, Graham, and Williamson in Contact Zones

Chemical laboratories—what a Satsuma student called *kemisuto dokoro* (chemists' place)—were undoubtedly the primary contact zone between

Japanese students and UCL chemists because they all started their student lives at UCL by taking a laboratory course of analytical chemistry.[32] Indeed, in 1864–1865 Graham, while still an advanced student, had been a classmate of three Chōshū students, Yamao Yōzō, Endō Kinsuke, and Inoue Masaru, in the course of analytical chemistry.[33] When Masaki Taizō, former retainer of the Chōshū domain and a close acquaintance of Inoue Kaoru, studied at UCL between 1872 and 1874, he lived with Graham and wrote to Inoue Kaoru that Graham "taught Inoue [Masaru] and Yamao earlier and is therefore friendly to Japanese."[34]

However, laboratories were by no means the only contact zone. Williamson not only accepted the Chōshū and Satsuma students in his chemistry classes, but he also had pastoral responsibility for the Chōshū students, finding households for them to live in, including his own, and teaching the three R's at home to those who lived with his family.[35] It is worth noting, however, that there seems to have been an invisible barrier between the Williamsons and Chōshū students within his typically middle-class Victorian household. Inoue Masaru, who stayed there the longest, until 1868, served the family as a student-servant (*gakuboku*) because of lack of additional funds from his domain.[36]

Later, for the student group of the Satsuma domain who came to Britain in 1865, Williamson played a different role as advisor and broker in addition to that of professor. He recommended two teachers of English reading (*dokushoshi*) and arranged for students to stay in separate houses of respectable teachers to make their practice of English conversation more effective after two months of staying in a common lodging together. Graham was recruited by Williamson for these purposes probably on the basis of his teaching experience with Chōshū students.

Mori Arinori, one of the Satsuma students, stayed with one of the teachers hired to help with teaching English. Mori called him "*Rigakushi* (Science Teacher or Master) Gurēmu," respected him highly, and apparently had a more amicable relationship with him than the one that prevailed between Chōshū students and Williamson, as illustrated by his photograph taken in Berwick-on-Tweed. He wrote on the back of the picture: "[I] stayed with him and often called him Father."[37] Mori's pictures suggest a third academic mentor. A picture of the other teacher of English for Satsuma students labeled "*Dokushoshi Bāfu*" and taken in London bears a signature of "Barff" on the back.[38] Hatakeyama Yoshinari, another Satsuma student, recorded the first name of this person in his diary, which can be read as "Frederick."[39] Thus, it is inferred that he was Frederick Settle Barff, an 1844 Cambridge graduate and clergyman who was then studying chemistry with Williamson at UCL.[40] Indeed, when Yoshida Kiyonari, another Satsuma student who later worked for the Meiji government as an expert in finance, was elected as the first ever Japanese FCS in 1872, he acquired recommendations based on "personal knowledge" from Graham and Barff as well as from Williamson and Foster whose classes he officially attended at UCL.[41]

In addition to their academic study at UCL and English practice at home, Satsuma and Chōshū students found it informative to observe the actual working of British technology in museums, dockyards, arsenals, mints, and several other kinds of factories. One such learning experience was an industrial tour organized by Williamson on July 29, 1865, to view the Britannia Ironworks in Bedford, which was renowned for the manufacture of state-of-the-art steam-powered agricultural machinery.[42] Though factually inaccurate, the "Visit of Japanese to Bedford" article in *The Times* does convey the excitement, joy, and enthusiasm this tour evoked among Japanese students for manufacturing and agricultural technologies, the machinery used, and the manual operation of the machinery as well as the British sense of novelty about an excursion of Japanese students to the factory.[43] Students' enthusiasm for agricultural machinery is easily understood, considering the Satsuma domain's policy of "enriching the nation" together with the landed interests of Satsuma high-ranking samurai. As I clarify in the next sections, this enthusiasm is a clue for understanding the students' attitude toward Graham's teaching.

Around the time of the excursion, Williamson was preparing for enrolment of Satsuma students in his laboratory course starting in September 1865. Graham was involved in the laboratory training of Satsuma students at UCL. Williamson petitioned the College Council for inclusion of the Japanese students in an abbreviated course of study:

> I should be glad to obtain the permission of the Council for a course of laboratory instruction of a somewhat exceptional kind which is now needed by a party of 14 Japanese Students. These young men cannot avail themselves of the full laboratory course but wish to enter the laboratory as Students for one year, working only 3 or 4 hours per day. I propose that a fee of fifteen guineas [£15,75] be charged for the 12 months course. No special arrangements will have to be made by the college for the Japanese of their instruction *as my assistants will do all that is needed*. The arrangement will not keep other students out of the laboratory by half-filling it as these young Japanese can work in the laboratory contiguous to the lecture room.[44]

As this quote shows, unlike the Chōshū students, the Satsuma students were trained not in the Birkbeck Laboratory, UCL's main student laboratory for practical chemistry, but in the preparation laboratory for demonstrations in the chemistry lecture hall. It was a sensible arrangement because this laboratory was most probably Graham's regular workplace at UCL. As Williamson was disabled in his left arm and right eye from his childhood,[45] his successive assistants, like Graham at the time, undoubtedly aided him greatly in preparing for his lecture demonstrations. The lab was also conveniently located next to the Birkbeck Laboratory,[46] where Graham had to monitor other students' progress.

The above narrative makes clear the different roles of Williamson and Graham in taking care of the Chōshū and Satsuma students in several contact zones. Williamson functioned as a supervisor of the whole educational enterprise, while Graham more closely interacted with both Chōshū and Satsuma students in several contact zones, such as laboratories, the joint lodging of Satsuma students, and his personal residence.

The "Graham Proposal" as a Window on Graham's Interaction with Students

What interaction, then, occurred between Graham and the Japanese students in the contact zones? In 1872, Graham wrote a direct account of his educational activities with Japanese students. After the Meiji Restoration in 1868, Itō, who had been a UCL student in 1863, returned to London in August 1872 as a deputy envoy of the Iwakura Mission, a high-profile delegation of the Meiji government to the United States and European countries with Iwakura Tomomi as the chief envoy.[47] Its main objective was to secure treaty revision and to study the West, but Itō also had other tasks, including investigating the situation of government-sponsored overseas students.

It sounds extraordinary that Itō consulted Williamson's laboratory assistant Graham, instead of Williamson himself about this national issue of government-sponsored overseas students, but this is what happened. It unequivocally shows Itō's recognition of Graham as *the* central figure in the contact zones between UCL teachers and Chōshū and Satsuma students. In his letter to the Meiji government in November 1872, Itō wrote:

> The accumulated harms [of the present system of government-sponsored overseas students] have become extreme, and the success of overseas students depends on luck. Very few of them will probably finish their studies successfully. Mr. Charles Graham, a London scholar [*Rondon Gakushi Charuresu Kurahamu shi*] taught Japanese students personally and deplores the above-mentioned harm. On the basis of his own experience, he has drafted a proposal, hoping that [the Meiji government] might reform its system of overseas study thoroughly for the benefit of the practical affairs of the nation [*kokka no jitsumu*]. Though his opinion might be a mere conjecture, his main point is the same as mine. So I have translated his draft and give it to you for consideration.[48]

As Itō mentioned in this letter, a Japanese translation of Graham's draft proposal ("The Graham Proposal") dated October 30, 1872, was enclosed.[49] It was circulated among the heads of government ministries and carefully read by the head of the Ministry of Education (*Monbukyō*), Ōki Takatō.[50]

In the light of these sources, it is obvious that Graham's acute criticism of the lack of proper selection and monitoring processes for students in the Japanese overseas study program, endorsed by Itō and carefully read by Ōki, contributed to the swift reform toward a meritocratic system based on competition within an emerging national school system controlled by a central

Ministry of Education. It is not a coincidence that one of Graham's former students, Masaki Taizō was appointed "superintendent of overseas students [*kaigai ryūgakusei kantoku*]" in London, a position created as part of this reform in 1876; Masaki Taizō had previously served as chemical laboratory assistant at the Tokyo Kaisei Gakkō for two years.[51]

However, the importance of the Graham Proposal lies not only in its immediate effect on the Japanese policy for overseas study. It consisted of Graham's own report of his teaching experience with Japanese students as well as his proposal for a government technological school developed on the basis of his experience. The proposal, therefore, reveals both his views on technological education and the events that occurred during Graham's interactions with Chōshu and Satsuma students in the various contact zones associated with their UCL experience. Graham started this proposal with a discussion of the aim of Japanese overseas study. According to his understanding, "there seem to me two things that the Japanese government wishes for this educational measure: first to transfer European arts (*gigei*) and science (*gakujutsu*) to educate its people and second to advance its wealth."[52]

Retrospectively, we can see that, in fact, Graham was calling for processes of transculturation, i.e., translation, selection, and modification of Western technology to suit Japanese domestic conditions as the British saw them. According to Graham, the combination of training in "arts" and "science" was the key to technological innovations in Japan because:

> Ingenious people are not generally content with emulation. They advance science beyond what other countries have done, invent new theories of machines, and innovate on age-old methods. It is only natural that new methods are better than old ones.[53]

However, his students did not share his perceived target of learning both "arts" and "science." Graham pointed out that, apart from tremendous linguistic difficulty, "the evil of hasty learning [*sōshin*]" hindered their progress:

> They want to acquire the substance of subjects for practical use [*jitsuyō gakka*] and to climb to the height all at once without sufficient training in the basis and outline of science, which should be regarded as the foundation of subjects for practical use. Without a basis in elementary subjects, they encounter difficulty when they start to learn subjects for practical use and want to learn what they should have learned beforehand, and so the evil of hasty learning and shortcuts increases. If they encounter difficulty in training in subjects for practical use, how can they select from what they learned and put it into practice by considering various circumstances when they are to establish these industries in their own country?[54]

To remedy this "evil," he proposed to Itō instituting a state school in Japan to train Japanese students thoroughly and prepare them to compete for overseas study.

Graham insisted that students in Japan should engage in learning the following core set of subjects regardless of their future career intentions, including: (1) Western languages; (2) mathematics and natural philosophy; (3) practical chemistry and physics; (4) geology, zoology, botany, and biology; and (4) drawing and other technological subjects [*kōgyō no gaku*]. Competition for overseas study should come after this stage, and selected students were to complete an apprenticeship at factories and workshops in Europe and America in order to be able to lead various industries in Japan afterward. In this sense, the school he proposed was primarily for technological education. However, as Graham emphasized, this school was not only for turning out leading technologists. Those who were not selected as overseas students would be "good enough to work in government offices all over Japan, to become teachers at elementary schools" or to be subordinates to leading technologists. In short, the Graham Proposal reveals a noticeable difference from his mentor's view on technological education. It represents the liberal science model, stressing broad science education as the basis for a variety of occupations, but it also illustrates an orientation toward a more specific training in technology to accommodate the needs of Japanese students.

I will discuss the impact of the Graham Proposal on the Japanese overseas study policy and contemporary development of higher education in Japan in the next chapter. Of particular importance here are his reflections on his teaching experiences with Japanese students. The proposal implies that conflicts arose in the contact zones in the UCL teaching environment. Graham, and presumably Williamson, insisted on the importance of purely scientific subjects while their students wanted to learn technological subjects from the outset. This conflict had a profound impact on the Chōshū and Satsuma students' views on science, technology, and education as well as on Graham's opinion.

Chōshū and Satsuma Students' Views on Science, Technology, and Education

The difficulty in analyzing the impact of Chōshū and Satsuma students' experiences at UCL on their views on science and technology comes largely from language and the image of Western learning in late-Tokugawa Japan. Domain students could use several Japanese words corresponding roughly to technology, such as *waza* (skill), *jutsu* (art), and *gei* (art). The Japanese words *gaku* and *gakumon* meant scholarly activities generally, largely based on the Confucian scholarly reading and interpreting Chinese classics as the primary model. The Japanese word for science used today, *kagaku*, was coined in the early Meiji period. Before then it was difficult to express ideas on the relationship between science and technology in a manner that translates easily to Western concepts.

Furthermore, the Western dichotomy between science and technology was itself unfamiliar to Japanese students. As the Chōshū slogan of "suppress the barbarians with the *arts* of barbarians" shows, Western learning was widely considered as arts (*jutsu*) by samurai who institutionalized and practiced it. As Satō Shōsuke emphasized, scholars of Western learning in late-Tokugawa Japan developed a rhetoric that underlined continuity rather than dichotomy between theory and practice.[55] Under these conditions, the Chōshū and Satsuma students tried to articulate the content of what they learned in Japanese, in English, and, most important, in what they did.

One index of the impact of the Chōshū and Satsuma students' experiences at UCL on themselves was the discrepancy between their intended subjects of study and what they actually learned at UCL. All Chōshū students took the laboratory course of analytical chemistry in the first year, and Yamao, Endō, and Inoue Masaru took lecture courses of chemistry, geology and mineralogy, mathematical physics, mathematics, and civil engineering in the following academic years in addition to their continuing enrolment in the laboratory course. All Satsuma students followed suit in taking analytical chemistry in their first year, and Mori, Hatakeyama, Yoshida, and Matsumura Junzō took "physical laboratory" in the following academic year.[56] Despite their initial conflict with Graham concerning their choice of subjects, they largely followed Graham's advice to study scientific subjects first, before technological subjects.

According to the record of a conversation in London between three Chōshū students and a British merchant with experience of living in Japan, the Chōshū students seem to have clearly articulated their mission in 1864. "They had been sent by their master, the Prince of Nagato [Chōshū], to Europe," wrote the merchant, "principally to study The Applied Sciences, and such arts as might be useful to their countrymen, and the English language."[57] An effect of their contact with Williamson and Graham is clear: their view on their mission changed from learning naval techniques to learning applied sciences and useful arts.

Itō and Inoue Kaoru, who had left for Japan shortly before, did not join this conversation, but they were well aware of what the remaining Chōshū students were doing. Itō wrote to Inoue Kaoru in a letter in May 1866: "Nomura [Inoue Masaru] is doing *bunseki seimitsu gaku* [exact science of analysis], and Yamao entered a dockyard in Scotland. I've heard that both are quite successful in their *gakugyō* [academic study]."[58] *Gaku*, with the adjective *seimitsu,* which means "exact and minute," should have meant something like "exact science" comprising precision measurements, something that was a hallmark of UCL's chemistry and physics education and of laboratory teaching in physics in Victorian Britain generally.[59] Itō (probably from Inoue Masaru) understood that what they studied at UCL was not technology itself but scientific enquiry involving precision measurements and techniques, and according to Williamson's version of the liberal science model, this would be applied later in their workplace.

The education of Inoue Masaru, whom Itō mentioned in his letter, illustrates this point. After finishing academic study at UCL with a Certificate of Honor in geology in 1866–1867, Inoue Masaru completed an apprenticeship in various mines before returning to Japan in 1868.[60] As a student-servant who lived with the Williamsons longest of all Chōshū students, he had reasons to "behave" and most faithfully followed Williamson's liberal science model. The academic achievements of the other Chōshū students are also worth mentioning. In 1864–1865, Yamao and Endō were awarded Certificates of Honor in analytical chemistry (fourth and fifth places, respectively), and Yamao was awarded a Certificate of Honor in chemistry lectures (tenth place). As noted above, their studies were supported by Graham.[61]

The Satsuma students embraced a slightly different image of science as applied to the useful arts. Hatakeyama used *gakujutsu*, a combination of *gaku* (scholarly activities) and *jutsu* (art, technique), to denote subjects such as chemistry, physics, and mathematics that he studied in the laboratories and at home.[62] Mori, who was more closely associated with Graham because he was also his boarder, more explicitly revealed Graham's impact. In letters to his brother written during his study in London, Mori described the object of his studies at UCL as "to learn and to receive training in one or two *gei no gaku* [science of art]." He later used *sho gei gaku* (science of several arts) to define his study area.[63]

Japanese students from the domains of Chōshū and Satsuma thus showed clear signs of accepting the liberal science model and the rhetoric of the applicability of science to technology. On the one hand, the Chōshū students, such as Yamao and Inoue Masaru who were most directly under Williamson's supervision, and (through them) Itō inclined toward Williamson's version of science education with more emphasis on "pure" chemistry. On the other hand, the Satsuma students, such as Mori and Hatakeyama, who had more contact with Graham than with Williamson, adopted Graham's version of the liberal science model incorporating *gei* and *gaku*. This difference might seem small at this stage, but it became noticeable in the process of institutionalization of chemical pedagogy in early Meiji Japan that was undertaken by both Chōshū and Satsuma students; I will discuss this in the next chapter.

Satsuma Students' American Study

Satsuma students including Hatakeyama, Mori, Yoshida, and Matsumura did not end their studies in London; they crossed the Atlantic in August 1867, and this further widened the difference between their experiences and those of the Chōshū students. This development foreshadowed the religious aspect of the Satsuma students' later experiences in the United States. Through their connection with the Japanophile British aristocrat, Lawrence Oliphant, they were persuaded by the British-born American spiritualist Thomas Lake Harris to cross the Atlantic to stay in the Brotherhood of the New Life, a religious community in Amenia (later moved to Brocton), New York, established by Harris.[64] After Hatakeyama, Yoshida, and Matsumura split from Harris because of his complete lack of understanding for their sense of duty toward the nascent nation, they moved to New Brunswick, New Jersey, and entered Rutgers College in 1868.[65] Of particular interest is the type of education

Hatakeyama, Yoshida, and Matsumura received at Rutgers and the influence their experience had on their views on science and technology.

They were all enrolled in the new Rutgers Scientific School, and between 1868 and 1869 they took a combination of scientific and literary courses, such as mathematics, chemistry, natural history, rhetoric, mental philosophy, modern languages, and composition and declamation.[66] Rutgers Scientific School had been established in September 1865 following the decision of the state of New Jersey to apply the Morrill Land-Grant Act to Rutgers College in 1864. This act provided federal lands and other forms of funding for states in the union to establish colleges for agricultural and technological education; the act is widely acknowledged by historians of American education and science as a galvanizer of the pragmatic tendency of American higher education in the post-Civil War period.[67]

Under the leadership of George Hammel Cook, professor of chemistry and natural science, and David Murray, professor of mathematics and natural philosophy, Rutgers Scientific School took shape. The school promised to deliver two courses, "civil engineering and mathematics" and "chemistry and agriculture," whose aim was to provide future engineers and agriculturists with the scientific basis of their occupations. Chemistry was to be taught with special reference to its application to agriculture.[68] In the early years, however, all students followed the same course due to a shortage of students.[69] Reflecting the history of Rutgers College as a classic liberal arts college[70] and the educational philosophy of Cook, Rutgers students were always trained in basic scientific principles and the liberal arts, as is shown by the subjects Hatakeyama, Matsumura, and Yoshida took.

The character of Rutgers Scientific School was thus similar in scope to the English liberal science model, but Rutgers put more explicit emphasis on the application of scientific principles, and in this resembled Graham's version of the liberal science model. There is a marked contrast to land-grant colleges in many other states where the professoriate responded to the criticism of farmers and introduced vocational training with subjects such as handicrafts and manual labor.[71] William H. Campbell, president of Rutgers College, vigorously defended Cook's vision:

> [The] Trustees have considered it no part of their duty to turn the agricultural department into a school of manual labor. They have from the beginning proceeded upon the theory that while the practical applications of science should be kept carefully in view in a course of instruction, yet that the main business of a scientific school must be to teach scientific principles and the methods of scientific investigation.[72]

The Satsuma students' perceived image of "science as applied to useful arts" shaped during their study at UCL was thus most likely consolidated by their later encounter with science education at Rutgers.

For Hatakeyama, his study in New Brunswick between 1868 and 1871 was important for other reasons as well. First, he was closely associated with one of the senior teachers at Rutgers, David Murray. In 1873 Murray was invited by the Meiji government to be superintendent of the Ministry of Education, a post he would hold until 1879.[73] As his trusted student, Hatakeyama supported Murray's work in Japan as the director of the Tokyo Kaisei Gakkō, the flagship institution of higher education under the control of the Ministry of Education; their cooperation continued until Hatakeyama's premature death in 1876. This close relationship between a Japanese student and a senior professor is a telling example of how differently the contact zones at Rutgers worked compared to how they had worked at UCL.

Second, as Inuzuka Takaaki has shown, Hatakeyama's growing conviction of the truth of Christianity led to his baptism at the Second Reformed Church in New Brunswick in 1870.[74] This turn is not surprising, because given the close tie between Rutgers College and the Dutch Reformed Church, Hatakeyama's wider collegiate life may have been influenced by Christian religious practices.[75] He thus formed the idea of his mission as educator including the task of enlightening the Japanese nation on the basis of Christianity; nevertheless, he still retained his earlier idea of "enriching the nation, strengthening the army" by "science as applied to useful arts."[76] As I discuss in the next chapter, these two aspects of Hatakeyama's student life in London and New Brunswick and his Anglo-American connections are the key to understanding his role as a "cultural mediator" between American and British teachers and Japanese officials and students during his directorship of the Tokyo Kaisei Gakkō.

Murray, Williamson, and Graham after Contact with Japanese Students

This chapter has highlighted how the Chōshū and Satsuma students' visions of science, technology, and education were affected by their encounter with Williamson and Graham and, in Satsuma's case, were strengthened by their experience of attending Rutgers College, a particular type of American land-grant college. One remaining question is whether and to what extent the interaction between UCL chemists or American teachers and their Japanese students was one-sided.

In the case of Murray, the impact of his association with Hatakeyama was so strong that it simply brought him to Japan and transformed Murray's career from that of a professor to that of an education administrator (see chapter 2). The case of Williamson is not as striking but still too important to ignore. In 1866, a year after the Japanese students' visit to Bedford's industry, Williamson began to organize similar industrial tours to chemical works for UCL chemistry students.[77] Japanese students' enthusiasm for machinery and viewing the actual working of technology, together with his experience in networking with manufacturers, arguably gave him the stimulus and opportunity to shape and implement his educational philosophy that chemical technology could only be taught in real workplaces for British UCL students. His contact with the Chōshū and Satsuma students informed his *A Plea for Pure Science* in 1870.

Graham's experience of teaching Japanese students, who were hungry for Western technological knowledge to improve their own nation's prosperity, had a more noticeable effect on his later career. After a three-year stint as analyst abroad, Graham resumed tutorial work at UCL in 1870 while making his career as a consulting chemist specializing in fermentation industries such as brewing.[78] He was appointed professor to the newly created chair of chemical technology at UCL in 1878 and kept this position until 1889; in the process he established himself as an important advocate of academic industrial chemistry in nineteenth-century Britain.[79] As he had barely started his career as a consulting chemist when he tutored Japanese students in the mid-1860s, it seems sensible to infer that his dialogue with Japanese students encouraged him to develop his own ideas of British technological education that shaped his later career.

Graham's inaugural lecture as professor of chemical technology in October 1879 suggests how much he owed to his experience of teaching Japanese students.[80] In this lecture, while supporting Williamson's liberal science model, he argued that without appropriate college training in "the application of pure science to industrial operations," it was extremely difficult for a young technologist to acquire technical knowledge and skills to improve existing processes on his own while also carrying out his daily duties in the factory. In this way, Graham justified the existence of his course of chemical technology at UCL, which specialized in the teaching of specific knowledge about various chemistry-related industrial processes and training in the "intelligent examination" of such processes. What Graham essentially did is to elaborate his earlier Graham Proposal for Japanese students based on his later experience as consulting chemist.

How then, did the Chōshū and Satsuma students use the ideas they had earned from this interaction? In the next chapter, I will show how the Satsuma students' vision of "science as applied to useful arts" took shape in the institutionalization of chemistry teaching in early Meiji Japan in the 1870s that was developed by British chemist Robert William Atkinson at the Tokyo Kaisei Gakkō and Tokyo University. Moreover, I will compare Atkinson's teaching program with that of another British chemist, Edward Divers, at the Imperial College of Engineering, Tokyo, the institution established by two Chōshū students, Itō Hirobumi and Yamao Yōzō.

Chapter 2

American and British Chemists and Lab-Based Chemical Education in Early Meiji Japan

With the Meiji Restoration in 1868 came two decades of a massive social experiment for a nascent nation-state to develop higher education as part of its industrialization-qua-Westernization policy. One key characteristic of this experiment was the predominant role of foreign teachers. In the hands of once rebellious samurai from Chōshū, Satsuma, and other domains that seized power, the formerly "illegal" enterprise of sending students abroad became the norm. Politician's enthusiasm for overseas study, reflecting their admiration as well as fear of Western civilizations, was matched by the sheer curiosity of the general public for the outside world even though overseas study was beyond the reach of most of the populace.[1] It was only natural for the new Meiji government to look for a more affordable option than simply sending students abroad. The underlying assumption shared by most Japanese in that period is clear: one could best learn about Western civilizations, science and technology in particular, directly from Westerners and by firsthand observation of them.

This simple description masks the sheer difficulty faced by students, teachers, and school administrators. Problems seemed endless. On the one hand, there were the issues of cross-national governance and communication. Who manages a school and with which managerial structure? Whom to hire as teachers from which country? Which language to use for teaching? On the other hand, there were issues specific to techno-scientific education. How should classrooms, laboratories, and other pedagogical settings be combined? And how to strike a proper balance between the scientific and technological components in designing a curriculum well suited to Japan in the early Meiji period? If these questions sound familiar to us living in the twenty-first century, this is because we haven't found a magic formula. It would then be hardly surprising that this early Meiji experiment was carried out in a hit-and-miss manner.[2] It is from this haphazard aspect that we can learn most about cross-national education in science and technology.

How did early Meiji educators meet such manifold challenges? In addressing this question this chapter analyzes one of the earliest cases of laboratory-based techno-scientific education in early Meiji Japan: the Department of Chemistry developed and run between 1874 and 1881 by British chemist Robert William Atkinson at the Tokyo Kaisei Gakkō, which became part of Tokyo University in 1877. It had its origin in a distinctly Japanese institution (I give a more precise meaning to the term afterward), a foreign-language school established by the Tokugawa Shogunate more than a decade before the Meiji Restoration in 1868. In the early Meiji period, the school had a mixed faculty consisting predominantly of American teachers and British latecomers and was headed first by a Dutch-American missionary and later by a Japanese director. This institution thus epitomizes the very messiness of the early Meiji pedagogical experiment. I first elucidate how the school, largely by trial and error, tried to solve the above questions and came to hire Atkinson as the first head of the Department of Chemistry. I will then look into how Atkinson used the legacy of his American predecessor, William Elliot Griffis, his own experience in London, and his interaction with his Japanese students to construct his style of chemistry teaching that favored applied chemistry by transferring and "transculturating" the educational models discussed in chapter 1 to Tokyo Kaisei Gakkō and Tokyo University.

Atkinson managed to set up a school of chemistry that proved influential in the development of higher education of science and technology in Japan, but his was not the only example. It was rivaled by the Department of Practical Chemistry at the Imperial College of Engineering, Tokyo, developed by another British chemist, Edward Divers; that department was oriented more toward the teaching of pure chemistry. The historiography of techno-scientific education in early Meiji Japan has long been dominated by this college as the most successful institution of its kind in that period.[3] Is this judgment warranted? I will intervene in this historiography by focusing more on Atkinson's teaching program and comparing it with that of Divers. I will argue that the question is not so much which was more successful but rather which different managerial and teaching styles were in play. I aim at highlighting the influence of different managerial and faculty structures, traditions (or lack thereof), and school cultures on the relationships between British chemists and Japanese school officials and students in different contact zones as well as on the actual contents and character of techno-scientific education there. Lastly, by looking at the collaboration between graduates from Tokyo University and the Imperial College of Engineering, I will suggest that the two pedagogical approaches are mutually complementary rather than exclusive.

Pre-Meiji Legacies

The Tokyo Kaisei Gakkō was not established de novo by the Meiji government. It had a long history with many antecedents, the most important of which was the Bansho Shirabesho, a product of the Tokugawa Shogunate's policy on Western learning in the Bakumatsu or late Tokugawa period. It is important to explore continuities and discontinuities between the Tokyo Kaisei Gakkō and its antecedents to identify its pre-Meiji legacies as they relate to institutional management, teaching methods, and the construction of contact zones.[4]

The arrival in 1853 in Uraga Bay (near Edo) of the squadron of Matthew Calbraith Perry, an US naval officer who had the mission of opening diplomatic relations with Japan, dragged the Shogunate into negotiations with the United States and other Western countries about commercial treaties in the late 1850s. The event also provoked serious concern about naval defense among senior Shogunate officials as well as interest in Western naval technologies. For these reasons the arrival of Perry has been singled out by historians of Japan as the starting point of the development of Japanese nationalism.[5] Moreover, a suddenly expanded requirement for the translation of diplomatic documents and the need for translation in the interest of defense technologies stimulated the Shogunate's commitment to the institutionalization of Western learning while retaining its long-standing anti-Christian policy.

The result was the establishment in 1856 of the Bansho Shirabesho (Institute for the Study of Barbarian Books), an appropriate name for a politically charged institution born out of "this intransigent imposition of foreignness on the Japanese body politic."[6] It is in this sense that I earlier characterized the Bansho Shirabesho a distinctively "Japanese" institution unlike other Shogunate schools in Nagasaki, which had a much longer history in foreign trades and interactions. The Bansho Shirabesho was renamed the Kaisei-jo in 1863, a name the school retained until the Meiji Restoration and partly passed on to the Tokyo Kaisei Gakkō.

Due in large part to the above circumstances in which the institute was established, the Tokugawa Shogunate assigned to the Bansho Shirabesho the following three functions: to translate diplomatic documents and Western books for Shogunate officials, to train translators and scholars of Western learning to meet its needs, and to censor Japanese translations of Western books to prevent the spread of Christianity. Therefore, the appointment of teacher-translators and the admission of students to the institute were made mainly on the basis of the Western languages students read, whether Dutch, English, French, German, or Russian. The third, censoring, function meant that Shogunate officials tightly controlled what was taught at the institute. The reading of Christian books was strictly forbidden, and the subjects taught were basically of a technological nature, seldom involving the humanities or social sciences. As Shogunate officials dealing with highly sensitive documents for national affairs, all teacher-translators were samurai either from the Shogunate or from domains and not from abroad.[7]

The choice of topics translated and taught at the Bansho Shirabesho strongly reflected the view of Western learning held by the Shogunate Confucian scholars who took control of the institute. They understood Western learning as a necessary evil, that is barbarians' *geigoto* (arts), or *geijutsu* (leaning and arts), and *hyakkō no gigei* (useful arts for various industries).[8] For this reason, special importance was attached to chemistry in the rhetoric of Shogunate officials and Confucians responsible for the institution's management. Chemistry was one of the first subjects given departmental status between 1860 and 1862 in the Seiren kata (literally "Department of Refining") to train specialists with technical knowledge.[9] In contrast, natural philosophy was largely suppressed because of its suspected connections with Christianity.[10]

Yet, laboratory teaching in the Seiren kata encountered considerable difficulties for lack of facilities, supervision, and a purpose-built laboratory, all problems acknowledged by Tsuji Shinji who studied and taught chemistry there and later became an official with the Ministry of Education.[11] Indeed, in 1867, for the purpose of launching laboratory-based chemistry and physics teaching, the Kaisei-jo decided for the first time to hire a foreigner, the Dutch medical doctor and chemist Koenraad Wolter Gratama. He had been teaching since 1866 at the Nagasaki Kaigun Denshūjo (Nagasaki Naval Academy), one of the pioneering institutions where the government-employed foreign teachers taught Western learning to Japanese students.[12] This project, however, was soon aborted amid political turmoil caused by the collapse of the Tokugawa Shogunate. Gratama was forced to move westward, and in 1869 he opened a school of chemistry in Osaka, the Osaka Seimi kyoku (later Osaka Rigakkō) whose significance in the development of science education in Japan, especially in the Osaka and Kyoto regions, has been well documented.[13] Nevertheless, the story of Gratama is instructive for my story in Tokyo (Edo) as well: Pedagogical difficulties in chemical education forced the reluctant Kaisei-jo to contemplate the creation of contact zones consisting of Japanese and a foreigner.

Bridging Past and Future: The Nankō in the Early 1870s

When the new Meiji government took over the Kaisei-jo in 1868 and reopened it in December 1869 as the Daigaku Nankō (which was renamed the Nankō the following year), it had inherited a "Japanese" institution of Western foreign learning whose pedagogy was dominated by the teaching of the art of translating and interpreting texts in small rooms and that was exclusively taught by Japanese teachers; the institution faced serious pedagogical difficulties, particularly in chemistry. Moreover, its management was tightly controlled by the government and the subjects taught were biased toward "useful arts for various manufactures."[14] The appointment of Dutch-American missionary Guido Fridolin Verbeck as head-teacher (*kyōtō*) in October 1870 marked the most important reform of teaching staff and method. He was entrusted with all aspects of school management, including the supervision of foreign teachers, mediating contractual negotiations between foreigners and Japanese officials, and drafting regulations and curricula.

Under Verbeck's directorship, the teaching method of the Nankō was rapidly transformed between 1870 and 1871 from *hensoku*, which meant the teaching of the art of translating Western books by Japanese teachers regardless of pronunciation, to *seisoku* defined as the teaching of all subjects in Western languages by Western teachers.[15] *Seisoku* was further divided into English, French, and German divisions according to the language used in lectures. Each section taught the general or preliminary course (*Futsū ka*), consisting of language, mathematics, world history, and natural sciences in its curriculum. As of January 1871, around 70 percent of all students chose English, 24 percent French, and 6 percent German.[16] This arrangement necessitated employment of a large number of English-, French-, and German-speaking teachers at the Nankō and marginalized Japanese teachers, who increasingly played an auxiliary role at this school. *Hensoku* teaching was abolished altogether in October 1871 for the perceived reasons of "its harmful effect of conservativeness [*injun no hei*] and the lack of prospects of success in this method."[17] A large number of contact zones in the form of classrooms thus emerged largely under the control of a foreign principal and foreign teachers.

Why did Verbeck assume such great power and responsibility in this school reform? Availability dominated the selection of foreign teachers at the Nankō during the first several years of the Meiji period, and Verbeck was certainly available.[18] He came to Nagasaki in 1859 as a missionary of the Dutch Reformed Church in America. After teaching at the Shogunate English School there, he became one of the first foreign teachers at the Nankō in 1869. However, most historians agree that he had several qualities that made him a natural choice as head-teacher of the Nankō. Born in Zeist near Utrecht in the Netherlands, Verbeck learned several languages in his youth including native Dutch, German, French, and English in addition to classical Greek, Latin, and Hebrew. He later got some command of Japanese after his arrival in Nagasaki, and he had an uncanny ability to establish rapport with Japanese students. He was widely read and had acquired almost encyclopedic knowledge, especially in law and politics. He also probably attended the Technical School in Utrecht before immigrating to the United States in 1852 to work as a civil engineer.[19] Verbeck's extraordinary abilities as linguist and educator suited the extraordinarily multilingual teaching environment of the Nankō.

It would then be tempting to consider the Nankō as Verbeck's school, treating him as a sort of king in a quasi-colonial manner. However, things were not as simple as that. The appointment of Verbeck and the radical departure from one legacy of the school's pre-Meiji predecessor—the pedagogy of Western learning based on translation—was made by the Ministry of Education, the new executive body of educational policy of the Meiji government. It was inaugurated in July 1871 soon after the abolition of domains, a major step of the Meiji government toward a centralized nation-state. The first head of the ministry (*monbu-kyō*), Ōki Takatō, an ex-samurai from the Saga domain and cousin of the reformist politician Ōkuma Shigenobu, was supported by his inner circles of former Kaisei-jo and Nankō teachers. They included Tsuji, Katō Hiroyuki (first president of Tokyo University), and Nakajima Nagamoto who has a role later in this story. Therefore, the ministry's scathing assessment of the traditional *hensoku* method, admitting "the lack of prospects of success in this method," has to be taken seriously as an insider testimony to the inherent limitation of teaching Western learning only by translation and interpretation of texts as was done at the Kaisei-jo. These former Kaisei-jo and Nankō teachers functioned as important bearers of institutional memory between the Kaisei-jo, the Nankō, and beyond.

The ministry's proposal of school reform to the Dajōkan (the central executive body of the Meiji government existed until 1885) in July 1871 shed more light on the thinking behind this reform:

> If we imitate foreign schools in all respects ranging from school structure to clothes, food, and rooms; if we select the cream of students from the Nankō, Tōkō [sister school of the Nankō specializing in medicine] and other schools and put them into dormitories; if we train and teach them thoroughly by the foreign method, and our students feel as though they were in foreign countries, then our students will find their way to learn according to their own abilities and acquire the general knowledge of *geijutsu* in Japan without going all the way abroad, which is so fashionable today.[20]

Unabashedly Westernizing on its surface, this quotation shows the ministry's (and therefore former Kaisei-jo teachers') painful recognition that scientific pedagogy did not exist in isolation but as part of the total experience of Western material culture. Indeed, glassware and other apparatuses for chemical experiments, the most relevant part of the material culture to scientific pedagogy, were initially totally lacking at the Nankō. This way of thinking was highly influential in the early Meiji period and affected many government-run schools, including what would later become the Imperial College of Engineering, Tokyo.[21]

Looking at this proposal more carefully reveals the ambivalence of the Ministry of Education officials. First of all, the proposal made clear that a competition with "fashionable" overseas study in the early 1870s made the "Westernization" of Nankō necessary. The emphasis was not on Westernization per se, but on returning the initiative of higher education into the hand of domestic school officials rather than to overseas study, which was then beyond their control. Indeed, the Ministry of Education's school reform to "imitate foreign schools" was combined with its decision, in October 1871, to dismiss eight of sixteen teachers at the Nankō as soon as their contracts expired because "they are all of especially low quality, and if we continue hiring them, they will affect the whole morale of the school amid the school reform."[22] While depending on foreign personnel, the ministry used fixed term contracts as a weapon to assert its authority and to assume tough control over school management.

Even Verbeck's appointment as *kyōtō* was only meant to be a temporary measure, as can be seen from a Nankō internal memorandum dated September 26, 1871:

> A *daigaku* [college or university] is a true *daigaku* only if it is equipped with specialist departments, and the Nankō has not yet reached this

> stage. Therefore we should still regard it as a language school, appoint Verbeck as head-foreign-teacher (*gaikoku kyōshi no kyōtō*) and for the time being entrust to him all decisions about school regulations and teaching methods. But, as his ability is limited, we will not regard him as a future head of the institution including specialist departments, but as head of the language department only.[23]

It is important to point out that almost from the outset the Nankō (and certainly Ministry of Education) officials started to think about reviving another of the pre-Meiji legacies, tight government control of the school, after Verbeck's term as head-teacher would end in September 1873. That would be when Japanese directors (*gakkōchō* or *gakuchō*) could assume total control over school management.

Equally significant in this quotation is a kind of road map, i.e., the Nankō evolving from a "language school," or what Robert Schwantes called "a replica of an American grammar school,"[24] to what ministry and school officials perceived as a "true *daigaku*" equipped with specialist departments. Just how strongly school officials craved for such a *daigaku* is apparent in the following episode.[25] Against its own good judgment, the Nankō attempted to establish a specialist college (*Senmon kō*) besides the Nankō in November 1871. For this purpose, in January the following year Verbeck was appointed a law professor (*hōgaku kōshi*). Griffis, an American teacher of physics and chemistry at the domain school in Fukui and a close friend of Verbeck, was appointed a science professor (*rigaku kōshi*). Perhaps predictably, however, this attempt was soon aborted due to the shortage of qualified students and teachers.

Interestingly, Griffis and Verbeck both used the term "polytechnic school" in their correspondence to refer to the specialist college.[26] They never attended, let alone graduated from or taught at, a polytechnic school, so it is a matter of informed guess to figure out what they intended. However, it would be safe to say two things: (1) they had American institutions in mind, otherwise mutual communication between Verbeck and Griffis would have been impossible; (2) a "polytechnic school" in the American context was understood as the combination of the original *école polytechnique*, an institution teaching basic scientific subjects to future technologists and engineers, and *écoles d'application* providing specialized training for civil engineers, mining engineers, etc.[27]

This is a telling discrepancy because, while the concept of "specialty" (*senmon*) or "specialist" (*senka*) per se seemed familiar to Japanese officials, the kind of pedagogy a polytechnic school embodied, namely, that "general studies" in basic science should come before specialized studies, did not.[28] This is exactly the same issue that Charles Graham had encountered in London in teaching Chōshū and Satsuma students (chapter 1). A question specific to Verbeck and Griffis is that of their credentials. Needless to say, Verbeck's credential as a law professor had a serious weakness. In comparison with Graham and Griffis's successors, Atkinson and Jewett, Griffis had a similar problem with his credential as an expert chemist that would become clear in the years to come.

At the moment Griffis's appointment at Tokyo was safe, but this turn of events put him in an awkward situation. Eventually he began to teach physics and chemistry and other subjects, such as history and English, to students of the preliminary course of the English division of the Nankō by April 1872 without modifying the original two-year contract.[29] This irregular arrangement was to cause much trouble to both Griffis and the Tokyo Kaisei Gakkō. However, the main point here is that later power struggles between them had deeper roots in the mindset of Ministry of Education officials who had wished to regain control of the school and to create a specialist college as soon as possible.

William Elliot Griffis: The Generalist Chemistry Teacher

Born in Philadelphia in 1843, Griffis entered Rutgers College in September 1865 when the college was in the process of reform incorporating more scientific subjects into its classical curriculum (chapter 1).[30] Griffis took mainly classical courses in his first and second years and finished his studies in 1869 with a bachelor of arts degree, not a bachelor of science. However, in his third and fourth years, he did attend science-related courses and was particularly impressed by the demonstration experiments in the chemistry class of George Hammel Cook, the professor of chemistry and natural sciences at Rutgers and a central figure in Rutgers Scientific School. After a year of teaching experience at Rutgers Grammar School, Griffis was recruited to take a position as physics and chemistry teacher in the Fukui domain through the connections between Verbeck, the Dutch Reformed Church in America, and Rutgers College. Given a strong missionary link between Griffis and Verbeck, the former's appointment at Tokyo was hardly surprising.

Griffis left little information about the chemistry teaching he delivered at the Nankō, but most historians agree that his teaching did not much change from that in Fukui, and that his main achievement at the Nankō was the introduction of Western-style science teaching with lectures using a blackboard and experimental demonstrations.[31] Griffis's "laboratory" plan in Fukui confirms this interpretation as it was actually another lecture room and/or preparation laboratory (next to the main lecture room) equipped with facilities for demonstration experiments.[32] His lecture notes were entitled "Chemistry and Natural Philosophy – An Outline of the Science of Chemistry" and "Chemistry – Qualitative Analysis," suggesting that Griffis treated chemistry not as a specialized subject, but as a general subject combined with natural philosophy. The teaching of natural philosophy, in combination with chemistry, entered classrooms at Tokyo under a new managerial environment but with next to nothing to equip Griffis's classrooms: he had to buy glassware and other chemical apparatuses anew in the treaty port of Yokohama.[33]

The textbooks Griffis used to compile his lecture notes were a mixture of British and American publications readily available in US cities on the East Coast, such as New York and Philadelphia. Among the most frequently used was Roscoe's elementary-level textbook, *Lessons in Elements of Chemistry* (New York, 1868), a more theoretically advanced college-level textbook by American chemist George Frederick Barker, *A Textbook of Elementary Chemistry* (Louisville, Ky., 1870), and Charles W. Eliot and Frank H. Storer's, *A Manual of Inorganic Chemistry* (New York, 1871), which was used for setting demonstration experiments.[34] Again, this confirms that Griffis intended to teach elementary and advanced-level chemistry as part of natural philosophy rather than as specialized subjects.

To control large classes of students of various ages and abilities, Griffis skillfully used the technique of dividing them into groups, teaching older students first, who in turn taught younger ones in small groups.[35] This pedagogical technique enabled him to adapt Western-style lectures to teaching in Japan, where students were accustomed to the traditional Japanese teaching method in small classes. He did so without changing the basic structure of his contact zone, i.e., of a lecture room where a lecturer imparts knowledge to students.

Also important is that his science teaching most likely had a religious overtone as part of natural theology. As he recalled later, "I [Griffis] thank my Creator every day that in his Providence I was early led to inquire into the constitution of the material universe" God created, and for him "certainly no branch of physical science seems more efficient to 'replenish and subdue the earth' than chemistry."[36] Separating science and Christianity, which Japanese education officials wanted to do, clearly did not make sense to him.

His students were particularly impressed by his simple but eye-catching demonstrations, which they saw as magic tricks.[37] As Hiraga Yoshitomi, one of students of Griffis at the Nankō who later became an industrial chemist, recalled:

> There was a teacher of general chemistry, called Griffis from America. He did lectures and demonstration experiments very skillfully. He did experiments such as: transforming water into ice by evaporating ether in hot summer; filling a cup with water, putting a granule of phosphorus in it, and introducing oxygen gas into water, which gave rise to flame in water; mixing powder of potassium chlorate with sugar; adding a drop of concentrated sulfuric acid which made it burn with violet flame;...I had never seen and heard of these kinds of things, and they caused a very strange feeling in me. I wondered if this was magic or *kirishitan bateren jutsu* [the art of Christian fathers] that I had heard of.[38]

It gives us a rare glimpse of the process through which the religious overtone of Griffis's teaching was coproduced by a teacher and students who associated their sense of awe with Christianity. Griffis's popularity among Japanese students in Tokyo clearly rested on such pedagogical skills to get his message through and on his adaptability to Japanese teaching environments without much interference from Japanese education officials.

This is not a trivial achievement. Difficulties arising from the scarcity of equipment for demonstration and the cross-cultural and bilingual nature of Griffis's pedagogical endeavor at Tokyo would have been paramount. Probably his proudest moment came on October 9, 1873, when three of his students gave demonstration lectures in front of Emperor Meiji and other luminaries on topics such as "burning phosphorus in water" (as Hiraga witnessed) and "preparing chrome yellow from sugar of lead and potassium dichromate and dyeing white cloth with it."[39] Lectures were carefully choreographed and supervised by Griffis as part of the inauguration ceremony of the Tokyo Kaisei Gakkō and its brand-new building. He clearly succeeded in training someone like himself, a lecturer-demonstrator. However, we might ask whether these methods would have worked equally well with more advanced students in a specialist course. For example, how would Griffis train students in quantitative as well as qualitative analyses that require different laboratory settings, different interaction dynamics in different contact zones, and different teaching techniques from those Griffis had been accustomed to? These questions would soon emerge along with Nankō's transformation from a preparatory school to a specialist college.

American or British? The Kaisei Gakkō and the Impact of the Iwakura Mission

The Nankō was renamed the Kaisei Gakkō and proclaimed its establishment as a specialist college in April 1873 ("Tokyo" was added to the school name in the same year, in August). Specialist or core courses (*hon-ka*) were actually not opened immediately, but it was decided that the Specialist College consisted of a Law School (*Hō Gakkō)*, a Science School (*Ri Gakkō)* and an Engineering School (*Kōgyō Gakkō*). Other important decisions were also made at this point. For example, as the English Division had the largest student population in the whole Kaisei Gakkō, the school chose English as the language of the core courses, citing school finance as the main reason.[40]

Then the selection of teachers for core courses, not from the United States but from Britain, and the construction of a new school building began. Connected to the building construction was the build-up of teaching materials such as glassware and chemicals, which were very likely of mixed origin, Dutch, German, American, and British. Around the same time, the Kaisei Gakkō arranged the transfer of experimental equipment from the Osaka Rigakkō (Osaka Seimi kyoku), directed by Gratama and his German successor Hermann Ritter, and the Shizuoka ken Gakkō (Gakumonsho) where a close friend of Griffis from Rutgers, Edward Warren Clark, was teaching physics and chemistry.[41] This measure was possible only after the abolition of domains and the ministry's aggressive centralization policy.

These are consequential changes that greatly affected the pedagogy of the Tokyo Kaisei Gakkō. They also invite several questions seldom asked by historians. For example, why did the Ministry of Education decide to proclaim the establishment of the Kaisei Gakkō as a specialist college at this time even though it could not yet institute specialist courses? Why did the ministry decided to hire British instead of American teachers for specialist courses? I argue that the answer to both questions is the impact of the Iwakura Mission between 1871 and 1873. Its official mission was to negotiate treaty revisions with the United States and European countries, but its major impact was as a massive "grand tour" of Japanese politicians and officials who observed the Western civilizations firsthand.[42]

As for the first question, a factor in the decision for a specialist college was undoubtedly that Itō Hirobumi, who was then the deputy envoy of the Iwakura Mission in Europe, addressed a letter in November 1872 to the head of the Ministry of Education, Ōki Takatō, with the Japanese translation of Charles Graham's draft proposal for a state school of technology based on Graham's version of the English liberal science model.[43] Indeed, Ōki in his reply said to Itō that he "carefully read Graham's draft four times" and that he was "pleased to see that Graham's proposal corresponds to the opinion of the Ministry of Education."[44]

What seems to have been particularly stimulating to the Ministry of Education is that Graham articulated a closely coordinated system of technological education, from language and purely scientific subjects to technological subjects, that would lead to overseas study. This would be a redefinition of the relationship between domestic higher education and overseas study. Earlier, as we have seen, the Nankō had been conceived as an alternative, not preparatory, to overseas study. Now, according to the recommendation of a British scholar experienced in teaching Japanese overseas students, the Ministry of Education would have to consider domestic higher education and overseas study two parts of the larger whole of technological education. A swift reform of overseas study policy by the Ministry of Education stimulated by Graham's proposal then accelerated the institutional development of the Nankō into the Kaisei Gakkō, which necessitated hiring teachers of advanced technological subjects. In the case of chemistry, such subjects should have been something like "applied chemistry" or "chemical technology" in addition to analytical training.

The Iwakura Mission also had an impact on the ministry's decision to invite specialist teachers not from the United States but from Britain. Nakajima Nagamoto, a senior official of the Ministry of Education and former Nankō teacher, came to prefer British to American education while he participated in the inspection tour of the Educational Commission of the Iwakura Mission to the United States and Europe. One of the major tasks of the commission was to select a superintendent of educational affairs (*gakkan*), a top foreign advisor to the Ministry of Education to replace Verbeck. In two letters dated in February 12, 1872, from Nakajima in Washington, D.C., to Ōki in Tokyo, Nakajima insisted on hiring a superintendent from

Britain rather than from the United States, arguing that "American education is more dominated by Christian theology than by the two subjects of physics and chemistry (*rika ni gaku*) and most scholars or university graduates here came from this school [of Christian theology]."[45] His negative assessment of physics and chemistry education in the United States was thus intertwined with his political concern with the spread of Christianity, which he deemed harmful. This had been a recurring theme in the school management of the Nankō and its antecedents since the pre-Meiji period.

Nakajima's statement is an interesting, if biased, observation of American higher education in the 1870s by a Japanese education official. From today's vantage point, it is easy to find flaws in his argument. For example, Nakajima had a particular type of college in mind, that is, a relatively visible and prestigious mainstream college education at Harvard and Yale, as is clear from the official report of the education envoy.[46] He failed to take into account recent developments of higher education in science and technology in the United States, such as the emergence of land-grant colleges in the 1860s.

This point was in effect acknowledged by his superiors with the appointment of David Murray, the professor of mathematics and astronomy at Rutgers College and a founder of the Rutgers Scientific School, as the superintendent of educational affairs in 1873. But the process was far from straightforward. The key to Murray's appointment was his promise as an organizer of national education policy. As Yoshiie Sadao pointed out, Murray's secular and utilitarian view of education that favored strong centralized state control was out of sync with mainstream pedagogical thoughts in the Unites States. But this view seems to have pleased Tanaka Fujimaro, the administrative head of the Ministry of Education (*monbu taifu*), and Kido Takayoshi, the deputy envoy of the Iwakura Mission in charge of military and educational affairs, who interviewed Murray.[47] Once he assumed the responsibility of selecting and hiring foreign teachers as part of his job description as superintendent, Murray was not necessarily bound to the preceding policy of hiring British teachers for specialist subjects.

Still, Nakajima's statement clearly indicates that the perceived inferiority of the American education in physics and chemistry existed in the minds of ministry officials, which in turn partly explains the ministry's decision to hire British teachers for specialist courses before Murray's appointment. Needless to say, this decision turned the future of Griffis in Japan upside down.

Conflicts, Disputes, and the Enduring Legacy of American Models

Griffis remained in the Kaisei Gakkō after its opening, resumed teaching in the preliminary course for future science and law students, and even participated in the official inauguration ceremony of the school building in October 1873. However, his name was not in its plan for the future. In a letter from the school to the Ministry of Education dated March 4,

1873, the school had indeed already asked the ministry to recruit teachers from Britain for the Schools of Law, Science, and Engineering through the Japanese legation in London.[48] The Meiji government sanctioned this proposal on June 22.[49] What this meant for Griffis became clear soon: On July 15 he received notice that his two-year contract expiring in early 1874 would not be renewed, which triggered a fierce dispute between Griffis, the Kaisei Gakkō, and the Ministry of Education.[50]

As Edward Beauchamp discussed, on the surface the bone of contention was the interpretation of Griffis's two-year contract with the Ministry of Education. As a counterargument to the school's interpretation, Griffis claimed that his contract was for his appointment at the Specialist College, which had just begun. As Beauchamp saw it, this was an incident of his "clash with Japanese bureaucracy."[51] However, it is not just the matter of bureaucratic interpretation because by deciding not to keep Griffis employed, the Ministry of Education cast doubt on his credentials to teach at the *senmon kō* (Japanese expression) or "polytechnic school" (Verbeck and Griffis's expression) for which he was originally hired. It is no coincidence that Griffis emphasized his "specialty" as well as his interpretation of his contract in his argument.[52]

Mutual distrust between Griffis and school officials was further compounded by the "Sunday question," that Kaisei Gakkō officials around the same time ordered foreign teachers to work on Sundays, which Griffis severely criticized.[53] For education officials, this development would have seemed to justify Nakajima's fear about American teachers. The timing could not be worse for Griffis: the Ministry of Education's and the Kaisei Gakkō's reassertion of their authority over foreign teachers occurred around the time when Verbeck had taken a leave and left Japan on July 10, 1873, before his formal resignation in September 1873. The school officials' order was part of their attempt to regain control of school management but resulted in confusion and conflict with foreign teachers.

Among other attempts at supporting his case, Griffis wrote a letter on October 13, 1873, to Tanaka and the directors of the Kaisei Gakkō. After reiterating his argument about the interpretation of his contract, Griffis moved to a different ground and argued that hiring a new chemistry teacher would not necessarily have to mean the dismissal of the existing teacher:

> Now, in all good American Colleges and Polytechnic Schools, there are three Professors of Chemistry, 1st General Chemistry, 2nd Metallurgy, Mineralogy and Mining, 3rd analytical and applied Chemistry. In the Kai Sei Gakko, two professors of Chemistry will be needed, my business is to teach General Chemistry. When the scholars are further advanced, Analytical and Applied Chemistry will be in order. If you were to ask me to teach Applied Chemistry, I should decline, because this is not my branch. If you wished me to remain three years, teaching Chemistry to the same classes, I should decline, for the scholars whom I now teach will be ready for a new professor of either the 2nd or 3rd Departments in Chemistry.[54]

This statement is revealing in several ways. First, Griffis argued that it was his "specialty" to teach general chemistry, a ploy that in all likelihood would have backfired. Perhaps unwittingly, he corroborated the Ministry of Education's implicit judgment that his ability as chemistry teacher was quite limited, and that he was thus unfit for a full-fledged specialist college. For the ministry it would have made more sense to ask a new chemistry professor from England to teach both junior and advanced students than to keep a disobedient teacher who admitted his unwillingness to teach applied chemistry and chemical analysis, the keystone of the professional training of chemists.

More important in Griffis's statement is his suggestion to the Ministry of Education and the Kaisei Gakkō about the future organization of chemical education. In essence, Griffis proposed adopting multiple professorships for a specialist department of chemistry, and that one professorship should be in analytical and applied chemistry, both based on his broad, if superficial, knowledge of chemical education in the United States. Sources of his information are not hard to find. Rutgers College and its Scientific School, his alma mater, obviously come first. Rensselaer Polytechnic Institute was where Cook, Griffis's chemistry teacher at Rutgers, was trained.[55] Columbia College School of Mines, which Griffis took as a model to design a chemical "laboratory" for lecture demonstration in Fukui, together with Rutgers and Royal Society of London, is also a strong candidate.[56] We could add other institutions such as Yale's Sheffield Scientific School and the Massachusetts Institute of Technology. In all these institutions, applied, industrial, or agricultural chemistry as well as metallurgy and mining were taught as an independent subject often in combination with analytical chemistry, in sharp contrast to chemical education in England in the same period.[57]

I will discuss the impact of Griffis's suggestion on the chemical education of Tokyo Kaisei Gakkō later together with other factors. Suffice it to say here that whatever outcome this would have would depend on the successful resolution of the dispute between Griffis and the ministry and school officials, a conflict that remained unresolved for six months.

Hatakeyama, Murray, and the Establishment of a Department of Chemistry

It was at this point that Hatakeyama Yoshinari, a former Satsuma student at UCL who had also studied at the Rutgers Scientific School, was appointed director (*gakkōchō*) of the Kaisei Gakkō on December 19, 1873. The task ahead of him was onerous, to say the least. According to the annual report of the Tokyo Kaisei Gakkō for 1874:

> The organization of the school was totally transformed in 1874. Before that, Mr. Verbeck, the head-teacher supervised foreign teachers and led teaching management. The director [*gakkōchō*] only controlled the administration.... From January of that year the director came

to control the employment of foreign teachers and take command of teaching management. That is why his workload doubled or tripled. That is the change in the office of director.[58]

What this passage did not mention is that these tasks would all start only after cleaning up the mess caused by the dispute between Griffis and the ministry and school officials.

Hatakeyama acted promptly. He met Griffis on the day of his appointment as director of the Tokyo Kaisei Gakkō.[59] Hatakeyama had converted to Christianity and was a friend of Griffis during his student years at Rutgers (chapter 1). If we believe Griffis, Hatakeyama with his pro-Christian attitude helped Griffis resolve the "Sunday question" in his favor and played a role in granting Griffis an extension of his professorship.[60] In this case, Griffis's good feeling was as relevant as wherever the reality lies. In fact, in extending his contract, Hatakeyama was not simply accepting his request. In his letter to the head of the Ministry of Education dated February 12, 1874, Hatakeyama proposed to continue Griffis's tenure for six months (instead of two years) because "the arrival of a teacher for the Science School whom the Ministry recruited from Britain is expected to be delayed."[61] Hatakeyama acted here as a shrewd school official for the benefit of his institution and yet with cultural sensitivity, extending Griffis's term at the Kaisei Gakkō to save his face only until the arrival from Britain of a specialist teacher.

As Hatakeyama's letter tells us, the two issues, resolving the Griffis dispute and securing specialist teachers replacing him, were closely connected, and the latter was not at all over at the time of his appointment. The Kaisei Gakkō first asked the Ministry of Education to request a teacher for the English-language Science School in March 1873 via the Japanese legation in London without specific instruction about his subject. Negotiations using diplomatic links, however, reached a deadlock, and the school had to use other connections. Hatakeyama had to collaborate closely with those involved in the selection of teachers from Britain.

Hatakeyama proved a perfect choice for this purpose. While in New Brunswick, he was a devoted student of David Murray who had been recruited as superintendent of educational affairs by the Meiji government since September 1873. He remained in this office until 1879, and one of his many duties was the selection and hiring of foreign teachers. Indeed, Hatakeyama had assisted Murray as interpreter and advisor since Murray's appointment.[62] Murray wrote first to Alexander William Williamson, professor of Chemistry at UCL, asking him "to recommend and send to Japan a Professor of Chemistry and a Professor of Technology (Engineering)."[63] As he received the letter in late February, Murray must have written this letter in January. The timing and addressee of Murray's letter makes sense only if we consider Hatakeyama's appointment in December 1873 and remember that he was a former student of Williamson and Murray who had not known Williamson personally.

As a result, Atkinson, who was a former student of and current assistant to Williamson at UCL, was selected as professor of "physics and chemistry" on Williamson's recommendation.[64] When he arrived in Japan in September 1874, it was the responsibility of Hatakeyama to discuss and finalize Atkinson's title as professor of "analytical and applied chemistry," his job description, and the name and character of his new department as the Department of Chemistry. Other appointments for the department ensued, albeit after Hatakeyama's death. The American chemist Frank Fanning Jewett was appointed professor of general and analytical chemistry between 1877 and 1880. The German geologist Edmund Nauman and the mining engineer Curt Adolph Netto formed a separate Department of Mineralogy and Mining in 1877, but they also taught geology, mineralogy, and metallurgy to chemistry students between 1876 and 1885.[65]

What informed Hatakeyama and Murray's decisions that largely determined the faculty lineup of the Department of Chemistry at Tokyo Kaisei Gakkō and Tokyo University? There are four possibilities worth considering: (1) the Graham proposal of a state technological school in November 1872; (2) Griffis' letter of October 13, 1873, detailing his suggestion of faculty composition that was in the hand of Murray by January 1874;[66] (3) Hatakeyama's learning experience with Williamson, Graham, and Murray; and (4) Murray's experience as the major architect of the Rutgers Scientific School system. The correspondence between Griffis's scheme and the actual faculty lineup seems striking, but existent sources do not allow historians to tell which factor was more relevant than others. Yet, this is not necessarily a serious problem, because these possibilities are not independent alternatives but interrelated factors that are largely congruent with each other. In particular, there is the suggestion of a specialist department explicitly tailored to the training of industrial chemists, based broadly on Anglo-American examples, but especially on the US chemical education in the post-Civil War era. Moreover, the Ministry of Education and school officials' also had a strong desire to establish a college to train specialists.

Hatakeyama's brief but successful tenure as director of the Tokyo Kaisei Gakkō is best characterized as that of a "cultural mediator." Robert H. Smith, professor of engineering there who came to Japan from Britain with Atkinson, stated that "education . . . in the University began to make great strides under the enlightened and influential guidance of Principal Hatakeyama, one of the most cultured and sympathetic of all Japanese."[67] The importance of Hatakeyama's Christianity in building rapport with the foreign faculty has been noted by historians.[68] However, as his mentor Murray recollected in his obituary of Hatakeyama, the latter's respect for Western culture and conviction "of the truth of the Christian religion" were mixed with patriotic sentiments, awareness of his origins as a high-ranking samurai of the Satsuma domain, and his "intimate knowledge of the traditions and institution of her educational system" that "fitted him in an eminent degree to aid in the reorganization of [Japan's] educational system."[69]

Murray's obituary of Hatakeyama can also be seen as testimony to Hatakeyama's two-way communication skills with English-speaking teachers. For Murray undoubtedly acquired his information about Hatakeyama's life and thought from their conversations, and they worked as a well-functioning team helping each other until Hatakeyama's premature death in 1876. This communication skill was another key to Hatakeyama's success as an influential director of the Tokyo Kaisei Gakkō. Hatakeyama could listen to the concerns of foreign teachers, but he could also make the Japanese view heard and present his arguments to both sides with sophistication. At Tokyo Kaisei Gakkō, the power relationship between Japanese officials and Western teachers was redefined by the Ministry of Education in 1873, when Japanese directors received authority to assume control over school management. However, this shift in power could work properly only in the hands of a skilful mediator and communicator like Hatakeyama and result in meaningful modifications of educational models by the Japanese.

London Model Transferred: Applied Chemist as "Consulting Analytical Chemist"

Like his superior Hatakeyama's, Atkinson's task ahead was overwhelming.[70] According to one of his students, Sakurai Jōji:

> As for lectures, we had inorganic and organic chemistry and chemical technology, metallurgy, and the history of chemistry. Concerning practical work, we had qualitative and quantitative analysis and assaying in addition to a general course of chemical experiments. When students would reach the third-year advanced course, problems were to be assigned to them; they were to do experimental researches about these problems. And their results were to be submitted as graduate theses. The teacher in the Department of Chemistry was the Englishman, Atkinson, an energetic and promising young scholar, who taught all subjects. Only Mr. Masaki Taizō assisted Atkinson, taking care of general chemical experiments and qualitative and quantitative analysis.[71]

That is, until the arrival of Jewett (1877), Naumann (1876), and Netto (1876), Atkinson was the only professor of the Department of Chemistry who was responsible for setting up the department, teaching all subjects, and supervising laboratory work of all chemistry students with only one assistant.[72] How could it be possible for him to meet such challenges?

Atkinson was born in 1850 in Newcastle-upon-Tyne in the north of England.[73] He studied chemistry and other scientific subjects at both UCL under Williamson and at the RSM under Edward Frankland between 1867 and 1872, when he earned by examination a first-class bachelor of science degree in chemistry from the University of London.[74] Chemical education at UCL and the RSM between the 1850s and 1870s was formulated on the liberal science model, which rested on a dichotomy of "pure" and "applied" chemistry and in which only the "pure" side of the subject could and should be taught at colleges (chapter 1). Indeed, the only publications by Atkinson during his assistantship to Williamson were highly theoretical, i.e., on the validity of the atomic theory.[75]

Atkinson could, therefore, draw on his broad learning and teaching experiences from London in his own teaching at Tokyo. The best source about his classroom teaching of analytical, inorganic, and organic chemistry is the notebook of (mostly) Atkinson's and Jewett's lectures taken by geology student Kotō Bunjirō.[76] It is outside my scope to examine this source in detail here, but it does show that Atkinson's lectures on organic chemistry made much greater use of structural formulas than Williamson's (see chapter 3).[77] Atkinson therefore got closer to the structure-centered teaching method of Roscoe and Schorlemmer at Owens College Manchester, the path his students Sakurai and Kuhara Mitsuru would pursue further in their own teaching (chapters 4 and 6).[78] Likewise, Jewett's chemical training at Yale's Sheffield Scientific School and the University of Göttingen with the renowned organic chemist, Friedrich Wöhler, prepared him well to teach junior students especially in qualitative and quantitative organic analysis and organic chemistry.[79] However, Atkinson had to develop single-handedly an independent course of applied chemistry for Japanese students.

Atkinson's solution to the problem of delivering lectures on chemical technology or "manufacturing chemistry" (*seizō kagaku*) was conventional.[80] With no prior experience of working in factories, Atkinson explained the scientific principles of well-known chemical industries, such as coal gas and alkali manufacture "by means of lectures and diagrams," supplemented by an industrial tour to chemical works. It was what most English chemists did as part of their lectures on chemistry, including Divers at the Imperial College of Engineering, Tokyo.[81]

A more successful part of Atkinson's curriculum was laboratory courses. Atkinson's first annual report submitted to the Tokyo Kaisei Gakkō commented on his second-year chemistry students in 1875:

> The second-year students of this department showed a remarkable aptitude for chemistry through this year's learning and began to do chemical investigations (*kagaku shiken*) on their own. From this I have to say that they take this science seriously and more and more aspire to study it.[82]

For analytical training of junior students, Atkinson and Jewett used standard textbooks such as Fresenius, Thorpe and Muir, and Thorpe.[83] For senior students, however, Atkinson responded to Japanese interest in the exploitation of natural resources by using samples of domestic natural products, such as milk, sugar, and iron ore in his basic analytical training.[84] He also assigned water analyses to his students to train their analytical skills, which resulted in published papers by Atkinson as well as by his students and partly contributed to the establishment of Japan's water supply system.[85]

Atkinson's laboratory teaching closely resembled the part of UCL's chemical education designed by one of his teachers, Charles Graham, for the training of consulting analytical chemists, where commercial goods, foodstuffs, and water from various sources were used as samples for chemical analysis.[86] Atkinson took the laboratory course of analytical chemistry between 1871 and 1872 at UCL and worked there as assistant to Williamson between 1872 and 1874, while Graham undertook tutorial work in the UCL Department of Chemistry as well as consultancy work, primarily with brewing businesses. Atkinson was therefore acquainted with Graham as both student and colleague at UCL. Together with Williamson, Graham was one of Atkinson's supporters "from personal knowledge" for his election to FCS in 1872.[87] On these grounds, I argue that Atkinson introduced Graham's laboratory teaching approach at UCL into his own at Tokyo.

It was a natural choice for Atkinson because Hatakeyama had also learned analytical chemistry from Graham during his study at UCL with other Satsuma students. Moreover, Atkinson's implementation of his laboratory teaching was made easier because his laboratory assistant at Tokyo until 1876, Masaki, had also been a student and boarder of Graham at UCL. In the case of water analysis, Atkinson's training at the RSM also possibly contributed to some extent, as his teacher there, Frankland, was famous for his consultancy in water analysis.[88] Anyhow, the first element of Tokyo's teaching program of applied chemistry was analytical training for consulting chemists. That was largely the result of Atkinson's effort to make the most of his range of expertise and idea of applied chemistry formed by his learning experiences in London for his new job in Tokyo.

Transculturation of the London Model: Applied Chemist as "Field-worker"

Atkinson's teaching program would end not at this point, but with a graduation assignment, whose purpose, according to him, was to "prepare students to improve the industries prospering in Japan," that is, Japan's traditional manufactures.[89]

A part of Atkinson's perspective shown in this phrase comes from his learning experience at UCL with Charles Graham. In the address in 1879 to a Japanese audience "On the Argument that Science and Practice Need Each Other in Chemistry" translated by his Japanese students, Atkinson mentioned Graham's article on the chemistry of brewing in arguing the applicability of chemistry to manufacturing industry.[90] Here Atkinson made clear that what he had learnt from Graham was his view as a consulting chemist on the applicability of the principles of chemistry to technology—that industrial processes might be improved and products refined by means of chemical analysis and other scientific investigations. Seen in this light, Atkinson's "improvement" project was derived in part from Graham. What was new in Atkinson's teaching program was that he chose traditional Japanese manufactures as the overall object of his project.

Atkinson integrated student excursions during summer vacation to traditional Japanese manufacturers into his course. Atkinson's students did fieldwork there and, using the analytical skills acquired in his laboratory course, embarked on laboratory work with collected samples under Atkinson's supervision. This combination of fieldwork and laboratory work under Atkinson's supervision resulted in graduate theses with titles such as "On Japanese Pigments," "On Japanese Tea and Tobacco," "On Japanese Wax and Vegetable Oil," "On Japanese Sweets," "On Soy Sauce," and "On Lacquer."[91] We can now consult three of these theses. The first two examples are "On Japanese Pigments" by Takamatsu Toyokichi and "On Shoyu" (soy sauce) by Isono Tokusaburō, both of whom graduated from Atkinson's Department of Chemistry in 1878. The third was "The Chemistry of Copper Smelting in Japan," written by Nakazawa Iwata, an 1879 graduate who became an assistant to Atkinson.[92] Atkinson's own research on sake brewing during his professorship at Tokyo University also originated from those student excursions.[93] It was significantly aided by Takamatsu and Nakazawa.[94]

Chemical analysis was, of course, an essential part of graduate projects assigned to Atkinson's students, just as Louis Pasteur's study of fermentation and the use of the microscope to investigate fermentation and degeneration processes of sake were vital for Atkinson.[95] He probably first learned about these techniques from Graham who explained these topics in some detail in his article on the chemistry of brewing before coming to Japan.[96] However, the fieldwork aspect of Atkinson's teaching program was not just the gathering and examination of raw materials and end products in terms of chemical analysis, as natural product chemists do, or what Morris-Suzuki has called "the great translation."[97] Rather, it is comparable to "participatory observation" that a cultural anthropologist does in his or her fieldwork, entailing a long stay in situ as well as intensive interaction with indigenous people to build rapport with them. Apart from the results of chemical analysis, the existing thesis copies and Atkinson's research paper on sake exhibit knowledge of indigenous manufactures and their technological details, as shown by their numerous diagrams that are available only through close collaboration with local manufacturers as informants.[98]

The comparison between cultural anthropologists and applied chemists in Meiji Japan may sound unusual, but it does highlight the essential role of Japanese students in the construction of Atkinson's teaching program of applied chemistry. Just as cultural anthropologists often need informants or collaborators from indigenous societies, Atkinson's project would simply not have been feasible without students' participation as mediators, interpreters, and practitioners. This comparison also explains why Takamatsu excelled in Atkinson's teaching scheme and his thesis received particular praise from him as "the model of a graduate thesis that can be published in extenso...and used as an important book for consultation in factories and expositions."[99] Most local manufacturers belonged to the same social class as Takamatsu; they were wealthy farmers who ran manufacturing businesses (see chapter 3). Takamatsu was therefore arguably well prepared for networking, communicating, and building rapport with such local manufacturers.

Atkinson's Japanese students also had recourse to what Sugimoto Isao called *jitsugaku*, Chinese and Japanese indigenous scholarly traditions including *honzōgaku* (the studies of herbal medicine), *nōgaku* (agriculture studies), and *bussangaku* or studies of local products. All of these were based on fieldwork in Chinese and Japanese localities and often resulted in encyclopedic reference works that were was made widely available by the development of publishing culture in Japan since the early eighteenth century.[100]

For example, at the end of his thesis, Takamatsu presented a comprehensive list of ten reference books, including two well-known Chinese encyclopedic books on *honzōgaku* and *gijutsugaku* (technology studies) published in the Ming period: *Bencao gangmu* by Li Shizhen and *Tiangong kaiwu* by Song Yingxing.[101] The list also included Japanese books from the mid-Tokugawa period in the same *jitsugaku* tradition with the same encyclopedic format: *Wakan sansai zue* (ca. 1713) edited by Terashima Ryōan, and *Nihon sankai meibutsu zue* (1754), an encyclopedia of Japanese local products and their manufacturing technologies edited by Hirase Tessai. *Keizai yōroku* (1827) by Satō Nobuhiro was a work on the art of ruling, with special emphasis on how to enhance the country's productive power by advancing manufacturing technology. *Bankin sugiwai bukuro* by Miyake Yarai was a more down-to-earth guide to a variety of local specialties for consumers in order that they could make informed decisions in purchasing local specialties. Isono used *Kōeki kokusankō* by Bakumatsu agriculturalist, Ōkura Nagatsune, and *Bankin sugiwai bukuro* for his study of soy sauce.[102] Nakazawa did not cite any references, but, considering the abundance of reference books on indigenous mining and metallurgical technology in the late Tokugawa period, it is more than unlikely that he did not use such sources in preparing his thesis.[103]

Atkinson was no exception: He used *Nihon sankai meisan zue* (1830) by Shitomi Kangetsu for an illustration of the interior of a sake brewery.[104] He was particularly impressed by the long-standing (Sino-)Japanese custom of *hi-ire* (heating) because it seemed very similar to and to have much predated pasteurization invented in the early 1860s:

> The student of science in Japan has a wide field before him; that system of isolation which has prevented the introduction of Western knowledge till within the last quarter of a century has not been entirely fruitless, for it has resulted in the development of industrial process which are as novel and interesting to the European as those of the latter are to Japanese. The scientific students of the university and colleges of Japan need not, therefore, look very far in order to find subjects that require investigation and explanation, and this search will, without doubt, add largely to the sum total of existing knowledge.[105]

While expressing needs for "investigation and explanation" by scientific methods, he showed a certain respect for indigenous manufactures in Japan.

As discussed above, Atkinson's phrase "to improve the industries prospering in Japan" originated partly in the discourse of UCL chemists who tried to show the industrial relevance of their expertise. Ultimately, the phrase might possibly date back to the ideology of material and moral "improvement" through science spread throughout Enlightenment Europe.[106] Colonial and postcolonial historians, moreover, would find parallels between Atkinson's phrase and what Michael Herzfeld called "crypto-colonialism," a hidden form of colonialism by means of cultural, economic, and epistemological hegemony.[107] This rings true to a certain extent: Atkinson's students used the same phrase of "improving" and applied chemico-analytical means to measure the "qualities" of products (see chapter 7). It is also worth noting that some of the perceived needs for "investigation and explanation" came from real commercial problems arising from global market pressure. The problem Atkinson tackled with sake was that, in spite of *hi-ire*, sake had a noticeably shorter shelf life than pasteurized wine.[108] Kuhara, one of Atkinson's students, investigated Japanese purple that had come out of use partly because it tends to fade easily (see chapter 3). Local manufacturers would likely not have supported Atkinson or his students' research if they had not recognized these problems. My point here, however, is that Atkinson's attitude toward Japanese manufactures was rather complicated and not the simple manifestation of a colonial mindset, with arrogance, admiration, and respect mixed together.

Atkinson thus learned the technicalities of Japanese indigenous manufactures from his Japanese students armed with references such as those listed above and directly from local manufacturers as much as his students learned chemistry from him. In this second fieldwork element of Atkinson's teaching program, therefore, Atkinson was a pupil and his students and local manufacturers were his teachers who brought Japanese scholarly and technological tradition into Atkinson's teaching.

Chemical Education, Institutional Structures, and School Cultures

Atkinson's approach was not the only way of teaching chemistry to Japanese students. As mentioned earlier, it was rivaled by the Department of Practical Chemistry at the Imperial College of Engineering, Tokyo, which had been developed by another British chemist, Edward Divers. He was trained at the RCC in London (Atkinson attended the same institute later, albeit with a different name) and Queen's College Galway in Ireland.

Two recent studies of Divers's school agree on the following points: In his teaching Divers focused primarily on the classroom teaching of facts and principles in chemistry and thorough training in chemical analysis in the laboratory; he encouraged students to do original research in pure chemistry based on their acquired analytical skills; he suppressed his students' wish to pursue the study of "practical applications of their knowledge"; and he was successful in substantially reducing the amount of students' training in factories or construction sites that officially constituted a large part of a student's life at the Imperial College of Engineering.[109] Furthermore, when Divers dealt with the chemical industry in his lectures, he was concerned mostly with the one-way transfer of Western-style industries, such as the alkali, cement, and soap industries, and seldom with traditional Japanese manufactures.[110]

In light of their common British educational background based on the liberal science model, the contrast between Atkinson's receptivity to Japanese industrial interests and Divers's relative indifference to them seems all the more striking. How did such a difference come about? Here I approach this question in terms of (1) the different institutional structures and (2) the school and student cultures of the two institutions.

By the time of the arrival of Atkinson, the Tokyo Kaisei Gakkō had established an institutional structure where foreign professors worked under the guidance of a Japanese director Hatakeyama, who were responsible not only for the administration of the school but also for the management of the teaching. As I discussed above, this is the result of a reversal from the earlier management style of controlling foreign teachers through a foreign principal. Among foreign teachers, Atkinson's higher status within the Department of Chemistry was recognizable in his higher salary compared to that of Jewett, but they were colleagues, not a boss and subordinate, in an US-style multi-chair professoriate.[111] The institutional structure of the Tokyo Kaisei Gakkō and Tokyo University was therefore both hierarchical (between Japanese directors and professors) and flat (between professors). Moreover, Atkinson's contract with the Ministry of Education was renewable every two years. He felt obliged to submit annual reports in Japanese, with the support of Japanese translators, to Japanese directors to show that he had fulfilled his contractual duties. Atkinson's more precarious position made his teaching more susceptible to modification from the Japanese side than Divers's.

In contrast, due in large part to Itō and Yamao's Westernizing policy, the Imperial College of Engineering never made such reversals throughout its existence. Its teaching was not directly controlled by Japanese, but by a British principal, first by Dyer and then by Divers. As heads of departments, college professors there were granted full authority over their teaching.[112] They were also granted longer-term contracts than Atkinson.[113] Practically, therefore, the predominantly British professoriate enjoyed more autonomy in teaching and was physically insulated from Japanese officials and, more generally, from the Japanese populace at large. According to a student recollections, Divers's knowledge of the Japanese language remained poor in spite of his long stay in Japan, which lasted until 1899; he made few friends with Japanese people other than his students and colleagues.[114] This disconnection was not limited to Divers's department. The case of telegraphic engineering examined by Graeme Gooday and Kakihara Yasushi shows that its teaching by William Ayrton at the college emphasized the pedagogical importance of electrical testing whereas most of his students were involved in telegraphic construction during the practical course, thus spoiling the coherence of Ayrton's teaching program.[115] In the case of Divers, such potential disjunction was avoided simply by the exemption from practical training in factories or on construction sites for chemistry students.

Furthermore, transculturation in the teaching and research activities of Atkinson should be viewed as a part of the wider issue of school culture at the Tokyo Kaisei Gakkō because it was the view of Japanese school administrators of the time that science teaching was part of the total experience of material culture, as we have seen earlier in this chapter. Science teaching thus exhibited a hybrid of Japanese and Western cultural elements. For example, the overall architectural style of the school's main building completed in 1873 (see figure 2.1) was at first glance Western. However, if one looks at details, side entrances were conspicuously roofed in the Japanese style, and visitors would not have missed the superbly designed Japanese-style front garden. Other architectural features of this building, which are difficult to depict in an illustration, include the wooden materials in the structure, the Japanese tiled roof, and the walls made of a traditional Japanese building material called *shikkui* (lime plaster), which was later covered with the quintessentially Anglo-American building material, wooden clapboards painted in white. This mixture of Western and Japanese styles is the main characteristic of what historians of Japanese architecture have called *gi-yōfū kenchiku* (architecture imitating the Western style).[116] A picture of Atkinson and his Japanese students just before their graduation in 1878 reveals a similar phenomenon. Western school uniforms were not introduced in the Tokyo Kaisei Gakkō; apart from Atkinson and a couple of students, most of those photographed wore the Japanese kimono.[117]

Figure 2.1. Bird's-eye view of the Tokyo Kaisei Gakkō. Courtesy of the National Diet Library, Japan

Source: *Takamatsu Toyokichi den*.

This hybridity in school culture indeed influenced the disposition of students there. Takamatsu Toyokichi, one of the first graduates from Atkinson's department, recalled:

> Students at the Tokyo Kaisei Gakkō regarded themselves as *bankara* in contrast to [those at the] Imperial College of Engineering, Tokyo in Toranomon. Prince Itō Hirobumi, who had just returned from Western countries, had a *haikara* opinion and dressed the students of the Imperial College of Engineering with Western-style school uniform and managed the school pretty elegantly. In contrast, students at the Tokyo Kaisei Gakkō had no school uniform and dressed very roughly like those before the Meiji Restoration. Even in graduation ceremonies they wore only *yukata* [informal summer kimono] with *hakama* [a long pleated skirt for men worn over a kimono] and received diplomas from officials roughly, as though they snatched them. [118]

By using the words *haikara* and *bankara*, Takamatsu contrasted the characters of students at the Tokyo Kaisei Gakkō and the Imperial College of Engineering, Tokyo. They basically represent the Western and Japanese styles, respectively, but also expressed the contrast between refined or deferential and rough or defiant manners, suggesting that power relationships between Japanese students and foreign teachers worked differently in the contact zones of these two institutions.

Takamatsu's recollection indicates that students in the classrooms and laboratories of the Tokyo Kaisei Gakkō respected but did not fear their teachers. Indeed, Atkinson's relationship with his Japanese students was apparently relaxed and friendly:

> Western teachers had cheerful and easygoing characters, and Mr. Atkinson often entered the classroom with *konpeitō* [Japanese confetti] in his mouth. They were very different from traditional rigid teachers, and we liked their characters very much. So we sometimes visited the private residences of Western teachers. [119]

This amicable student-teacher relationship arguably led to a fruitful collaboration between Atkinson and his Japanese students at the Tokyo Kaisei Gakkō.

Divers's relative indifference about traditional indigenous manufactures was in fact in conformity with the overall character of the Imperial College of Engineering, Tokyo and its parent Ministry of Public Works, which were concerned with the transplantation of British technology into Japan. This was also the case in all cultural aspects of student life at the college. The college building was superbly built in the Western style (see figure 2.2), and students there lived in the thoroughly *haikara* style:[120] they resided in its residence hall built in the same style, wore Western-style uniforms, and ate Western food such as beefsteak and stew. Playing cricket and soccer

Figure 2.2. The former Imperial College of Engineering, Tokyo (*Kōbu Daigakkō*), with former professors and students in school uniform, June 1893. Courtesy of the National Diet Library, Japan
Source: *Takamatsu Toyokichi den.*

were their favorite pastimes.[121] Japanese students of the Imperial College of Engineering were deliberately isolated with power relationships favoring a stronger British influence, which is markedly different from the situation at the Tokyo Kaisei Gakkō and Tokyo University in the same period.

Pedagogical Outputs and Impact on Industry

It has now become clear that Atkinson's and Divers's chemistry teaching in Tokyo had different styles and objects embedded in different contact zones, institutional structures, and cultures. That makes it difficult, if not meaningless, to assess who was more successful in terms of their impact on chemistry and the chemical industry in Japan. We cannot conclude, for example, that Divers's teaching was a failure because it did not include sufficient industrial topics. The same applies to the possible criticism that Atkinson's teaching was not "chemical" enough. To avoid such pitfalls, we must turn to the assessment of their main pedagogical "outputs," i.e., students, both qualitatively and quantitatively.

During Atkinson's professorship, 26 students graduated from Tokyo University between 1877 and 1881. It is perhaps unsurprising that a majority of them, namely 18, acquired positions in the teaching sector either at the secondary or tertiary level. It is noteworthy, however, that six graduates became analysts at the Geological Survey of the Ministry of Agriculture and Commerce (*Nōshōmushō Chishitsu Shikenjo*), a precursor of the Industrial Research Laboratory (*Kōgyō Shikenjo*) in Tokyo established in 1900.[122]

The most notable of them was Takayama Jintarō (grad. 1878), a friend of Takamatsu's who assumed the first directorship of the Industrial Research Laboratory in 1900. Both institutes in their early years were centers of excellence in industrial chemical research in Japan, which was concerned with "the great translation" of scientific research into traditional techniques such as the manufacture of lacquer, porcelain, dyestuffs, and paper.[123] The impact of Atkinson's teaching program on the development of the chemical industry in Meiji Japan is unquestionable.

During Divers's professorship, 25 students graduated from his department between 1879 and 1888 (note that Divers's tenure was longer than that of Atkinson).[124] In spite of the effort of Wada Masanori and myself, comprehensive data on their careers is still lacking. This is an obstacle in gauging the department's impact on the chemical industry in Japan. What Wada found is that a majority of the graduates whose subsequent careers are known were hired by national and municipal government ministries and agencies outside the Ministry of Public Works.[125] In this sense, the example of its most famous graduate, Takamine Jōkichi, of Taka diastase and adrenalin fame, is also a typical one; he took up a position as engineer (*gishi*) at the Ministry of Agriculture and Commerce. This example gives us a glimpse of Divers's impact on students.[126]

Takamine was one of the first six graduates from Divers's department in 1879 and surely the brightest of them: He was sent to Britain for overseas study between 1880 and 1882 by the Ministry of Public Works with students from other departments. In his first year at the Anderson's Institution in Glasgow, Takamine collaborated with British chemist Edmund J. Mills in his highly analytical research on the absorption of weak reagents by cotton, silk, and wool. The research was published later in the *Journal of the Chemical Society: Transactions* with Mills and Takamine as coauthors, which is a good indicator of Takamine's solid analytical skills honed in Divers's laboratory.[127]

On the one hand, it is unquestionable that Divers's rigorous analytical training was relevant to Takamine's subsequent career as one of the most successful industrial chemists in Japan and the United States. This also applies to others graduates who found positions in a printing bureau, mines, mint, petroleum, sugar, and other chemical companies, in army and navy arsenals, and in colleges.[128] On the other hand, Divers's teaching of technological chemistry focused on Western technologies seemed to be out of sync with Takamine. When choosing his entry-level position as an engineer at the Ministry of Agriculture and Commerce in 1883, he is said to have explained the reason as follows: "If you want to establish an industry developed in Western countries, it is best to hire Westerners who are well versed in its technologies. . . . I would like to apply the science I mastered in the most meaningful way, i.e., to Japanese indigenous industries. . . . To do that, I have to investigate first the current situation of such industries."[129] This is exactly the leading idea behind Atkinson's, but not Divers's, chemistry teaching. Unsurprisingly, therefore, when Takamine launched one of his first important R&D projects, the extraction of fungi with strong fermenting power from sake malt (*kōji*) for the production of whiskey, he chose an alumnus of Atkinson's department at Tokyo, Hida (or Hita) Mitsuzō (grad. 1879) as his collaborator.[130]

Takamine indeed "mastered the science" largely through Divers's pedagogy, but his aspirations as to how and for what purposes to use that knowledge and skill came from elsewhere. The collaboration between Takamine and Hida epitomizes the complementary characters of Atkinson's and Diver's teaching philosophy and of the school cultures of Tokyo Kaisei Gakkō and Imperial College of Engineering.

Chapter 3

The Making of Japanese Chemists in Japan, Britain, and the United States

Government schools in early Meiji Japan, including Tokyo Kaisei Gakkō and Tokyo University, are often dubbed "ex-samurai's schools" (*shizoku gakkō*). In quantitative terms, this label certainly rings true: According to Amano Ikuo's estimate, a whopping 70 percent of those graduating from Tokyo University in 1885 were ex-samurai though only 5 to 6 percent of the whole population in early Meiji Japan was from this class. The same was true for about 72 percent of students at the Imperial College of Engineering in the same year.[1] These numbers give us the impression that the student bodies of these schools were homogenous, an assumption that has given rise to the discussions of "samurai" characteristics in Japanese science and technology.[2]

However, a large number of students from one social class does not necessarily lead to a bigger impact on the subsequent developments of science and technology. As explained below, a variety of "filtering" mechanisms worked against some former samurai students during the training process in the school and beyond. In addition, social categorization sometimes hides important differences between people classified as belonging to the same group. These two factors make the consideration of students' personal circumstances important for evaluating the impact of an educational institution qualitatively. This chapter examines the background and academic development of four early Tokyo Kaisei Gakkō students who would themselves become key players in academic chemistry in Japan after completing their studies abroad: Matsui Naokichi who later graduated from the Columbia College School of Mines with a Ph.B. and Ph.D., Sakurai Jōji who completed his studies at UCL, Takamatsu Toyokichi who moved to Owens College Manchester and also spent time at the University of Berlin, and Kuhara Mitsuru who completed his Ph.D. at the JHU.

After they returned to Japan, all four students took up professorships of chemistry at Tokyo University. However, their approaches to scientific and technological education differed significantly, and this had a tangible impact on the institutionalization of chemistry teaching in Japan. I explore their early lives, their student lives at Tokyo Kaisei Gakkō and Tokyo University, and their later overseas studies for possible origins of their different views on chemistry teaching. This chapter therefore both supplements chapter 2 and serves as a bridge to later chapters.

Students' Early Lives, *Kōshinsei*, and Other Entrance Routes

It is useful to take up the parallel case of the "first generation" of Japanese physicists in the early Meiji period as a starting point of this discussion. That generation was characterized as a group of boys born in the 1850s, all from samurai families, who were "trained to the age of about 15 in the Chinese [Confucian] classics, and then were chosen to pursue Western studies either in Japan or abroad." This was no mere coincidence: They were typically selected by domains from among those who finished traditional samurai education at domain schools around the age of 13 or 14 and were then sent to Tokyo according to the *kōshinsei* system of the Nankō (the antecedent of the Tokyo Kaisei Gakkō, see chapter 2) set up in 1870.[3] *Kōshinsei* literally means students offered to Emperor Meiji by domain lords. This early measure to create a student body for the Nankō in a short period of time essentially rooted in the *ancien régime* ensured a high percentage of students who were former samurai.

The *kōshinsei* system also ensured the agony of many selected students. A majority of the Nankō students, many of whom were *kōshinsei*, suffered from the lack of experience in Western learning, especially languages, and had to quit the Nankō during its temporary closure in 1871.[4] Language problems were not the only ones faced by ex-samurai students. As Ronal Dore argued, unlike the Western three R's, the elementary education for samurai was based on Chinese Confucian classics and emphasized not acquiring knowledge and skill for daily life and material prosperity but rather nurturing the ethical values of piety, loyalty, and self-discipline. These nurtured values would then ensure that, as the ruling class, the samurai could govern the nation by virtuous behavior.[5] As a result, many samurai students at the Nankō and its successor institutions felt an "inner conflict" between traditional Confucian education and Western learning. For example, Tanakadate Aikitsu, a professor of physics at Tokyo Imperial University and a colleague of Sakurai, had difficulty abandoning the old Confucian maxim of "learning is for governing." During his preliminary studies at Tokyo University, he sought in vain "Western books explaining how to train oneself and govern the nation that measure up to the teachings of Confucius and Mencius."[6] To what extent, then, do these characterizations apply to our sample of chemists?

The *kōshinsei* system and traditional samurai education affected some chemistry students. The most conspicuous was Sugiura Shigetake, the chemist who became a prominent nationalist educator.[7] A *kōshinsei* from the Zeze domain (part of today's Shiga Prefecture), he chose chemistry as his major because he regarded it as an auxiliary subject for agriculture, and he believed that agriculture was the basis for governing in accordance with his Confucian outlook. Sugiura's wish to study "agriculture at the heart of the nation" was well known to his fellow students and teachers at Tokyo and influenced his choice to study at the Royal Agricultural College in Cirencester, England.[8] Another *kōshinsei*, Matsui completed his early education in Chinese classics in the Ōgaki domain (now part of the Gifu Prefecture), and his upbringing as a cultured samurai may have affected his later professional activities, which went far beyond his expertise as chemist. He had heavy administrative duties as a college dean and president of Tokyo Imperial University and as a top official in the Ministry of Education. He was even on the jury of a fine art competition.[9]

Yet there are two major problems in this line of arguments. First, the *kōshinsei* system did require selected students to have samurai status, but it did not necessarily determine their family occupations. Kuhara, who was selected a *kōshinsei* by the Tsuyama domain (now part of the Okayama Prefecture), is a case in point. His father was a Dutch-style medical doctor (*ranpōi*) serving the domain lord.[10] As Western learning was part of his "family business," he was well prepared in Classical Chinese, Dutch, and English language training (though more in reading and writing than in speaking and listening) when he entered the Nankō. From the outset he was firmly committed to learning Western science, especially as it pertained to medicine.[11]

Even more important, the *kōshinsei* system was soon abolished a year after it was implemented; it was replaced by entrance examinations based mainly on written and oral tests in a Western language.[12] Sakurai was one of the first students who entered the Nankō in 1871 under the new system. He was a former samurai from the Kaga domain (today's Ishikawa Prefecture) but was too young as a *kōshinsei*. This substantially affected his educational background: His elementary education as a samurai was not school-based, and it was brief, taking place between 1866 and 1869 with private tutors; then he entered a domain school of English in Kanazawa (the domain capital) in 1870 on the advice of his mother.[13] Moreover, during his year at the school of English in Kanazawa, Sakurai was sent with other pupils to Nanao, the remote naval port of the Kaga domain on the Noto Peninsula, to learn English for seven months directly from an Englishman without an interpreter.[14] Sakurai thus turned to Western learning at the relatively young age of 11 and acquired strong command in spoken English before entering the Nankō. He could easily pass an entrance examination and adapt himself to the teaching environment of the Nankō, where all teaching was delivered by Western teachers in their languages.[15] Having had relatively little exposure to traditional education, he was also much more open-minded about Western learning and culture in general and showed no sign of any intellectual struggle comparable to that of Tanakadate.

Takamatsu, who was also admitted through entrance examinations, affords a wholly new perspective on the early educational backgrounds of Tokyo Kaisei Gakkō students. He was born into the family of a wealthy farmer who held the hereditary position of *nanushi* (village head) in Asakusa, Edo (Tokyo).[16] As most sons of well-to-do commoners did in the Tokugawa period, he learned Japanese calligraphy and reading at a school for commoners called *terakoya*. He later pursued classical Chinese studies and arithmetic with a samurai teacher who was a shogunate official at the observatory (*tenmon-dai*) until he became a probationary village head in 1866.[17] After the abolition of the *nanushi* in 1869, Takamatsu began to teach himself English, wishing to study a specialist subject of Western learning. Takamatsu was admitted to the Nankō in 1871 at age 19.

Takamatsu's early education differed from that of most samurai youth most of all in emphasis and motivation. Among the samurai class, arithmetic was long considered a craft of merchants and was hardly a mainstream subject in a samurai's education. It was either omitted (as in Sakurai's case) or limited to a minimum.[18] In contrast, Takamatsu's study with a samurai teacher at the shogunate's observatory was centered on arithmetic, in accordance with his parents' conviction that "everyone has to be versed in numerical principles (*sūri*) regardless of his station" and that "all enterprises (*jigyō*) are inseparable from numerical principles."[19] There is little doubt that acquiring financial and business sense for his future role as a *nanushi*, whose duty was to administer the village as its representative and to give wealth to villagers by advancing industry, was at least part of what Takamatsu's parents meant. This factor is important for understanding his later role at Tokyo Imperial University as a trainer of chemical technologists for private chemico-industrial companies from the 1880s on.

By these examples we can see a variety of routes, motivations, and preparation with which youngsters in early Meiji Japan were drawn to education in chemistry at the Tokyo Kaisei Gakkō. It surely belies the image of their alma mater as an "ex-samurai's school." If the student body thus created was not monolithic, their experiences at school could not be either. How can we understand their variations in a structured way?

Age, Overseas Study, and Students' Experiences at the Tokyo Kaisei Gakkō

Let's start with the consideration of students' ages, which we all know have a tremendous impact on learning experience. We have two extreme cases: Sakurai and Takamatsu. Sakurai studied at Tokyo between 1871 (when he was 13) and 1876 (when he was 18). He attended the English division of the preliminary course for three years and the specialist course of the Department of Chemistry for two years before traveling to London as a government overseas student. On the other hand, Takamatsu studied there between 1871 (when he was 19) and 1878 (when he was 26). He attended the preliminary course for four years and the specialist course for three years before graduating from Tokyo University. It is evident that Takamatsu attended the school at a much later age than Sakurai.

This likely has had two effects on Takamatsu's and Sakurai's learning and their relationships with Atkinson. First, Takamatsu's age most probably made it more difficult for him to adapt linguistically to the teaching there by Western teachers in English. This is reflected in the fact that Takamatsu took a longer time to finish the preparatory course than Sakurai. Takamatsu himself candidly recalled the linguistic problems, which persisted after he entered the Department of Chemistry. He thus talked about the existence of "sound teachers," that is, interpreters, in the classes of the preliminary course and students' secret intrusions into the professor's office at night to peruse reference books so as to fill the gaps in their notebooks.[20] Sakurai also recalled that Takamatsu several times borrowed his notebooks taken in Atkinson's class.[21] As Sakurai was at the top of Atkinson's chemistry class in July 1875, it is clear why Takamatsu singled out Sakurai to ask him for help.[22] Takamatsu could have asked Matsui as well, another young student famous among his countrymen for his remarkable memory, if he had not been selected as a government overseas student and travelled to New York City in 1875.[23]

However, it is worth remembering that it was Sakurai and Matsui who had been unusually young. Age factors may have accelerated their learning, but it seems to have taken an emotional toll, at least on Sakurai. He recalled that he had to immerse himself in his studies during his five years in Tokyo in order to compete with *kōshinsei* students, who were mostly at least two or three years older than he was.[24] The young Sakurai therefore had little time to do otherwise and may have felt difficulty connecting himself to other older students. Indeed, a contemporary informal and rather gossipy source about the personalities of students of the Tokyo Kaisei Gakkō suggests that Sakurai was seen as diligent (*benkyō-ka*) but as not particularly interesting by his fellow students.[25] This is in contrast with Takamatsu and *kōshinsei* students such as Kuhara; both of them were known to have enjoyed many hobbies in their student years, including baseball, and to have had the humorous habit of tricking fellow students by impersonating others.[26]

Comparing Sakurai, Takamatsu, and Kuhara leads us to the second age factor, that is, the "relaxing" effect on the relationship between Atkinson and his students. When Takamatsu entered the Department of Chemistry in 1875 at the age of 23, Atkinson was 25 years old, i.e., roughly the same age as Takamatsu. Kuhara was younger (19) but not as young as Sakurai (17). Takamatsu's relationship with Atkinson was relaxed and friendly, and Takamatsu liked Atkinson's cheerful and easygoing character, which was different from that of "traditional rigid teachers," albeit without losing respect for him as teacher (see chapter 2). Kuhara's employment as junior assistant (*jun-jokyō*) upon Atkinson's recommendation to support his teaching and to continue Kuhara's own studies a year after his graduation in 1877 not only testifies to his ability but also suggests that Atkinson found him easy to work with.[27] The fact that they were of a similar age likely played a role in this amicable and nonauthoritarian relationship between Takamatsu, Kuhara, and other chemistry students and Atkinson as teacher and research supervisor.

Another important factor affecting students' experiences at Tokyo was the timing of overseas study in their respective careers. Matsui left Tokyo in 1875 to do overseas study in the United States after only one year with Atkinson. Sakurai was selected as a government overseas student by the Ministry of Education and went to England in 1876 before entering the final, third year of the Tokyo Kaisei Gakkō. They both missed doing a graduate thesis under Atkinson's supervision, a fact Atkinson deplored.[28] Kuhara and Takamatsu were among the first students to complete Atkinson's entire teaching program, and their graduate research was supervised by him. Kuhara was granted the degree of *Rigakushi* (bachelor of science) by the newly established Tokyo University in 1877 based on his thesis on "Japanese dyeing and printing methods," and Takamatsu received his degree in 1878 for his thesis entitled "On Japanese Pigments."[29]

Atkinson ranked Takamatsu's thesis "On Japanese Pigments" at the top of the ones submitted in his year and praised it highly in his annual report to the university as "the model of a graduate thesis that can be published *in extenso*...and can be used as an important book for consultation in factories and expositions."[30] The main objectives of his graduate thesis, as Takamatsu explained in its preamble, were "to study these compounds [Japanese pigments] as practically as possible and to give analytical results in addition to their history, methods of preparation, uses &c."[31] Though Takamatsu could not improve on or make innovations to existing processes in his thesis, he succeeded in meeting his objectives in the following 47 pages of his thesis.

Not only did Takamatsu record the results of his analyses of Japanese pigments, he also described the processes and apparatus for their preparation by drawings with detailed information about their dimensions. These were based on his own observations and interviews during fieldwork excursions with local manufacturers as well as on reference books in Chinese and Japanese.[32] These sources belonged to the indigenous scholarly tradition of *jitsugaku*, which were themselves based on fieldwork on local manufactures (chapter 2). He also paid attention to economic aspects of the industry by listing localities and prices of products for some of the pigments.[33] Most of the analytical work was done by Takamatsu himself, but the large scope of his work, covering 29 pigments, including nine pigments that were enamel colors for porcelain, necessitated some kind of collaboration with fellow students, of whom Takamatsu acknowledged three.[34] These features—the perseverance shown in his chemical analyses, vast knowledge of indigenous manufactures, the meticulous observation of technological details, and the close collaboration with local manufacturers and fellow students—were presumably the qualities that Atkinson had in his mind in praising Takamatsu's thesis.

Unfortunately, only the title of Kuhara's thesis is known, and there is no direct record showing how Atkinson evaluated his thesis. However, there is little doubt that Atkinson saw his work in a positive light. Kuhara's first article was published in *The Journal of the Chemical Society: Transactions* in 1879 as the first of the series "Contributions from the Laboratory of the University of Tôkiô, Japan" that also included Atkinson's own.[35] Kuhara's paper was concerned with the isolation and elemental analysis of a red coloring matter (and its bromine and chlorine derivatives as well as barium salt) extracted from a plant root (*Lithospermum erythrorhizon*, called *shikon* in Japanese). Most likely it was a part (or developed part) of Kuhara's graduate thesis on Japanese dyeing and printing and sent to London by Atkinson as worthy of presenting at a society meeting and of possible publication.

Kuhara approached a similar topic as Takamatsu's from a very different angle: he concentrated on the chemistry of one single substance and completely omitted the detail of its manufacturing process, which had already lost its economic competitiveness. According to Kuhara's own explanation, it had been used in Japan as purple dye but "on account of the fugitive character of this colour, and from the recent introduction of aniline colours, its use has been almost entirely abandoned."[36] Instead, he stood firmly on the methodology of structural organic chemistry, preparing the salts and derivatives of a target compound for analytical purposes, that is, "with a view to ascertain the nature of this colouring matter."[37] Indeed, this project would later turn into the graduate thesis topic of a student of him with the objective of determining the structure of this substance (chapter 4). His motive of taking up this topic is not clear, but most likely he chose it both for the excitement of puzzle-solving and for "improving the industries prospering in Japan" along the lines Atkinson delineated in his class report to Tokyo University (chapter 2).

Takamatsu's and Kuhara's graduate researches therefore testify to the breadth of Atkinson's capacity as research supervisor as well as to his students' different motivations, abilities, and interests. There is one similarity between Takamatsu's and Kuhara's studies at Tokyo, though: For both their studies served as a basis for their overseas studies and subsequent careers, Kuhara's as an organic chemist and Takamatsu's as a technological chemist, more than was the case with Matsui's or Sakurai's careers. In other words, the timing of departure affected the relative importance of domestic education compared to overseas study.

One question should be examined, though, before discussing individual cases of overseas study, namely, why the first group of government overseas students went to the United States in 1875 and the second group to Britain in 1876? There was a certain rivalry among students at the Tokyo Kaisei Gakkō. A series of reforms of the government-sponsored overseas study system were carried out by the Ministry of Education between 1873 and 1875 (see chapters 1 and 2). These reforms introduced more rigorous selection criteria, based on examination performance, and were coordinated with the establishment of the Specialist Course at the Tokyo Kaisei Gakkō. Students in the most advanced classes there "sold" themselves vigorously as candidates for the new type of government-sponsored overseas study trip in order to complete their studies.

The atmosphere was highly competitive as the first and second groups of overseas students were selected from the same classes by the Ministry of Education. The first set of overseas students from the English-language departments of chemistry, law, and engineering all went to the United States in 1875, a natural choice given that the advisor to the Ministry of

Education was American and the students' memory of American teachers was still vivid. As the biographies of Sakurai's fellow students Sugiura and Hozumi Nobushige indicate, the rivalry at the school induced some students left behind to contact school and Ministry of Education officials, expressing a desire to go to Britain in order to surpass the first group of overseas students.[38] Because all teachers in the English-language departments of law, chemistry, and engineering at the Tokyo Kaisei Gakkō in this period were hired from Britain, there was a conviction among students in the 1870s that Britain, not the United States, was the leader in Western learning especially in these three fields. As the following case studies amply show, it is difficult, if not pointless, to assess which countries offered students a better education, mainly due to institutional and intradisciplinary differences: each school discussed below had its own strength and weakness. This episode shows how perceptions affect the course of history.

Matsui at Columbia's School of Mines: Analytical and "Armchair" Industrial Chemistry

Matsui, one of the first overseas students selected from the Department of Chemistry at the Tokyo Kaisei Gakkō, enrolled in the course in analytical and applied chemistry at Columbia College School of Mines. Established in 1864, this school experienced a phenomenal growth in tandem with the rapid development of the chemical, engineering, and mining industries in the Gilded Age in the United States.[39] He entered the school together with his fellow chemistry students, Nanbu Kyūgo and Hasegawa Yoshinosuke who enrolled in the course of mining engineering. Columbia's School of Mines was what his chemistry teacher William Elliot Griffis had in mind in designing his lecture rooms in Fukui (chapter 2). Having spent just one year with English chemist Atkinson, Matsui was perhaps more under the spell of Griffis than any of the other chemistry students at Tokyo.

Because Griffis, Murray, and Hatakeyama had worked at Tokyo Kaisei Gakkō, there were certain similarities between the curricula at Tokyo and Columbia. As the course title implies, the Columbia curriculum was heavily biased toward applied chemistry and included a large amount of lectures on this subject in addition to lectures on inorganic chemistry, organic chemistry, geology, mineralogy, and metallurgy. There were also drawing practice and lab trainings in qualitative, quantitative, and blowpipe analysis and assaying.[40] The similarities between Columbia and Tokyo did not stop there. During summer vacation students who finished the first and second year had to do vacation work on a topic assigned by the faculty and then had to write an essay based on their work, followed by a graduate thesis, much as in Atkinson's teaching program.[41] Exactly to what extent the curriculum at Tokyo was based on information from Columbia is hard to tell. It is unlikely that Columbia's direct influence, if any, went beyond Atkinson and other professors' job titles (chapter 2). In any case, Matsui would not have found a huge gap between what he experienced in both schools. The same point would be made in 1879 by Kuhara who visited Columbia College when he stopped by en route from Tokyo to Baltimore.

Similarities between the curricula at Tokyo and Columbia in applied or industrial chemistry more likely came from their common origin in the professional activities of consulting analytical chemists. The professor of analytical and applied chemistry and the first long-time dean of the faculty at Columbia's School of Mines was Charles Frederick Chandler, one of the founders of the American Chemical Society. Trained at Harvard's Lawrence Scientific School and the University of Göttingen, he was one of America's premier industrial and public health chemists and a driving force behind the growth of the School of Mines.[42] A difference, however, is that, whereas Atkinson himself was relatively inexperienced in such work before coming to Japan, Chandler was a seasoned consulting chemist and had served both the chemical industry and municipal authorities. His consultancy therefore more directly affected the contents and style of his lectures.

A student's notebook of Chandler's lecture entitled "Industrial Chemistry" delivered later in 1902–1903 is filled with his favorite topics as consulting chemist such as pollution of air, water, and soil by microbes, dust and lead poisoning; boiler incrustation in railway industries; sewage; manures and fertilizers; and urban sanitation.[43] As these lecture notes cover only one semester (his lectures on applied chemistry were supposed to run over two years, i.e., four semesters), they do not show the whole range of Chandler's lecture topics.[44] What they do show is that Chandler skillfully combined official lecture topics, such as "air," "water," and "bacteria," with worldly themes drawn from his experience as consultant.

How did Matsui react to this pedagogical regime? Chandler himself wrote a grade report in 1876 for Matsui, Nanbu, and Hasegawa to Mekata Tanetarō, an official at the Ministry of Education who accompanied the overseas students from Tokyo Kaisei Gakkō to the United States as their superintendent. He praised their accomplishments, especially in analytical chemistry, adding that for this reason he omitted their examination in this subject.[45] Interestingly, after being awarded a Ph.B. in 1878, Matsui temporarily moved to Yale's Sheffield Scientific School in New Haven, Connecticut, to take a biology course before starting his postgraduate study back in New York.[46] It is not clear why he took this action, but a plausible explanation would be that Chandler's course included quite a few biological, especially bacteriological, components, which inspired or forced Matsui to supplement his knowledge in biology.

More revealing is the essay Matsui submitted to the School of Mines in November 1877, entitled "Memoir on Enamelling of Iron Wares."[47] This is clearly intended as a writing assignment ("memoir") based on vacation work after his second year in summer 1877; its title closely matches one of the topics specified for vacation work of the students of analytical and applied chemistry in that year.[48] The essay is unusual, to put it mildly. Though school regulations required vacation work to be based on "the personal examination of works in actual operation" and explicitly stipulated that students should mention "works visited and books consulted," Matsui did not mention any works he might have visited. Or he could not do so, because most of the content of his essay essentially came from two reference works: Crookes and Röhrig's *A Practical Treatise of Metallurgy* (1869) and Ure's *A Dictionary of Arts, Manufactures, and Mines* (1856).[49]

Another essay on the same topic entitled "Memoir on Enamels for Iron Ware" written by Matsui's classmate James Atkins Noyes is in better shape.[50] Under the section "Books consulted" he cited 23 references, including Crookes and Röhrig, Ure, the lecture notes of Columbia's professor of mineralogy and metallurgy, Thomas Egleston, and even shop catalogues for enameled goods.[51] However, Noyes did not mention any works he might have visited, either. This raises the question whether Matsui's and Noyes' behavior was an aberration from a norm or a more or less usual practice. Was there a hidden school regulation for students of analytical and applied chemistry?

To answer this question, it would be necessary to identify and examine all essays written under this scheme, which is outside the scope of my study here. However, one can conclude that Matsui's and Noyes's behavior was not that dissimilar from Chandler's own method as a consulting chemist. According to Margaret Rossiter, who examined the manuscript collection of Chandler, it consisted of perusing and clipping from encyclopedias, textbooks, journals, newspapers, and trade journals to familiarize himself with the technical details of a wide range of applied chemistry. That is, once a relevant case appeared, Chandler could focus on "the finer points, the particular tests, and the precise and advantageous definitions so necessary to a successful legal case," and he usually delegated the chemical analysis to his assistants.[52] We could then understand that by writing an essay Matsui was learning the first "armchair" phase of industrial chemistry. The other assigned topics could also be written up in a library without visiting a works, for example, the essays "A Scheme for the Qualitative Analysis of Type Metal, containing Pb, Sb, Sn, As, and Cu," "The Technology of Petroleum," and "The Chemical Nature of Meat and of Flesh Extracts."[53] The extent of Matsui's personal interaction with Chandler cannot be ascertained based on sources, but he likely assimilated his teacher's methods and ideas about the industrial chemist's expertise directly or indirectly, consciously or unconsciously.

Matsui's doctoral dissertation is of a totally different caliber than his school essays.[54] It was concerned with the chemical analysis of raw materials for the Japanese porcelain wares *Arita-yaki* and was published in the *Journal of the American Chemical Society* in 1880. His choice of journal is important as it indicates Chandler's approval of Matsui's work. Matsui first clearly formulated a research question, i.e., to which extent "the raw materials used in the manufacture of porcelain in Japan and China are somewhat different from those chosen for the same purpose in Europe." Then he disclosed the sources of his samples, a large manufacturing company in Arita, explained how he used power tools for mining, such as the Blaker's crusher to crush samples into fine powders, and proceeded to the presentation of chemical analysis of 16 samples from the body, glaze, thin layer between the body and the glaze, and pigment of the ware.[55] He then concluded that the "peculiarity of the Japanese porcelain consists in the relatively large amount of SiO_2, together with the small percentage of Al_2O_3, while CaO is present only in a very small quantity."[56] Matsui's paper constituted part of the robust field of analytical chemistry of Japanese, Chinese, and European porcelains in the late nineteenth century, a respectable outcome and proof of his completion of training as an analytical chemist.[57]

In spite of its much higher quality, however, Matsui's dissertation exhibits the same characteristics as his school essay—the lack of involvement with actual manufacturing processes or factory operations. Samples were provided by a manufacturing company, and Matsui's use of power tools for mining did not mean that he was involved with mining operations. On the contrary, it shows his approach to technology as a tool for analytical research. This noncommittal attitude to industrial on-site operations set Matsui apart from Atkinson and especially from Takamatsu, as discussed later in this chapter. Matsui's unpublished essay tells us that it was a product of Columbia's applied chemistry education.

Sakurai at UCL: "Physicalist" and Theoretical Approaches to Chemistry

Sakurai's student years at UCL between 1876 (when he was 18) and 1881 (when he was 23) defined his chemical career, just as Matsui's New York years defined his. Sakurai did well in his studies in London. He won the Gold Medal in Williamson's junior class of chemistry in 1876–1877, was granted the Clothworkers' Company Scholarship for 1879 based on a joint examination of the senior classes of physics and chemistry in 1877–1878, and began independent research under the supervision of Williamson. He was elected a FCS in 1879 and presented a paper on organomercury compounds at the 1880 BAAS Annual Meeting in Swansea, Wales.[58] This was published in toto in the *Journal of the Chemical Society: Transactions* in the same year.[59]

In contrast to Chandler's teaching at Columbia, Williamson's chemistry teaching at UCL displayed key elements of the liberal science model (see chapter 1). Indeed, Williamson's lectures were designed to train students' minds in scientific reasoning. In his only published textbook, *Chemistry for Students* (1865, second edition 1868; third edition 1873), Williamson adopted an inductive way of presenting his subject matter. It was designed for the use of beginners and students who "wish to have an outline of the chief facts and theories of mineral and of organic chemistry," describing and comparing "individual facts, so as to lead the mind of the reader toward general principles, instead of stating the general principles first and then proceeding to illustrate them by details."[60]

According to the two-volume lecture notes taken by William Dobinson Halliburton in 1877–1878, Williamson's chemistry course basically followed this system but was theoretically more demanding than his published textbook.[61] Both in inorganic and organic chemistry, the description of preparations, properties, and reactions of elements and their compounds came first. The fact that the notation of chemical formulas was only briefly explained in tandem with these descriptions suggests that he expected students to familiarize themselves with chemical notation elsewhere, possibly in UCL's chemistry exercise class, which he introduced with his former assistant, Henry Enfield Roscoe, in the 1850s.[62] Williamson laid out the elements to be studied according to their properties, i.e., nonmetallic and metallic elements. Metallic elements were classified according to whether they were precipitated in the presence of grouping reagents used in qualitative analysis, such as hydrogen sulfide, ammonium sulfide, and ammonia. In organic chemistry, the compounds to be studied were classified according to functional groups, i.e., alcohols, ethers, acids, aldehydes, and ketones.

Williamson explained chemical theories on this descriptive basis and, in turn, used such theories to review and classify experimental facts. For example, the concept of combining power (valency) was thoroughly explained after description of the four elements of oxygen, hydrogen, nitrogen, and carbon, and the general discussion of atomic theory came after he finished the description of all the nonmetallic elements that his lectures covered. In organic chemistry, the concept of types was introduced in his explanation of the etherification of alcohols in the presence of sulfuric acid, for the clarification of which in terms of water type he was already renowned as a chemist. Possibly for educational reasons, Williamson avoided graphic structural formulas with bond signs throughout his textbook, and in his lectures deferred their use until it became absolutely necessary, for example, in the explanation of the different properties of two isomers of propyl alcohol by structural formulas.

The major deviation of Williamson's lectures from his textbook was that chemical physics occupied the first two months of his lectures but was totally absent from his textbook. These notes show that Williamson's lectures on chemical physics amounted to a brief introduction to theoretical and experimental physics, comprising precision measurement of heat, light, electricity, and magnetism.[63] This part of Williamson's lectures had much in common with Division B (experimental physics) of the physics course taught by George Carey Foster as recorded in Halliburton's two-volume lecture notes on physics, whereas Division A of Foster's course was devoted largely to mathematical treatment of mechanics.[64] In short, in the hands of Williamson the liberal science model took a particular form of teaching physics as the basis of chemistry.

This was embedded in the curriculum of the Faculty of Science at UCL that Williamson helped frame. First, both chemistry and physics were required for the London degree of bachelor of science since its implementation in 1860.[65] Second, science students at UCL who wanted some kind of scholarship had to turn to the Clothworkers' Company Exhibition, which required them to excel in both chemistry and physics. Instituted in 1875, this exhibition was funded by the Clothworkers' Company, but the choice of subjects was made by the UCL Senate. Though Williamson was not a member of the internal committee of the senate that discussed the scheme for the exhibition, it was on Williamson's initiative that the senate appointed this committee, and he most probably exerted some influence through other committee members, such as Thomas Hewitt Key, his father-in-law, and George Carey Foster, his former student and assistant who had become his physics colleague in 1865.[66] Sakurai did not aim at a London degree, for which he would have been ineligible on grounds of lacking the specific general education requirements for the matriculation examination, but he chose to seriously study chemistry and physics with Williamson, Foster, and Oliver Lodge, demonstrator in physics at UCL, in order to sit the joint examination for the Clothworkers' Exhibition.[67]

Both "physicalist" and theoretical approaches to chemistry greatly influenced Sakurai. Sakurai included an introduction to experimental and theoretical physics in his chemical lectures and later even separated out that section of his lectures as an independent course on theoretical and physical chemistry.[68] Sakurai's training at UCL in precision measurement of physical properties such as temperature and electricity would prove to be a key element of his later laboratory teaching as well as his own researches at the crossroads of organic and physical chemistry.

In a letter to Williamson's daughter, expressing his indebtedness to her father, Sakurai commented on Williamson's greatness as a theoretical chemist and on his systematic approach to chemistry and chemical education:

> Nor was he [Williamson] much of an experimental chemist; in fact, owing to his bad eyesight, he was, I think, obliged to give up all experimental work in later years. But, as a theoretical chemist, he was one of the greatest men of his time.... It was, more particularly, Dr. W[illamson]'s discovery of the so-called "mixed ethers" that gave a final blow to the Equivalent System. There was nothing particularly wonderful about the experimental part of this ever memorable work of Dr. W[illiamson]'s, but it was the keenness of his insight that was so remarkable and the logic of his argument that was so convincing.
>
> As a lecturer, Dr. W[illiamson] was always most clear, precise and impressive, and I always took such a delight in hearing him. In fact, I think I got my habit of endeavouring at clearness and precision from attending lectures.[69]

In chapters 4, 5, and 6 I discuss more in detail how the above-mentioned features of Williamson's lectures affected Sakurai's views on the teaching of chemistry.

Williamson carefully supervised Sakurai's research, looking after his professional development as well as his scientific work. In his debut paper, Sakurai expressed his gratitude to his mentor, saying "I cannot conclude these notes without expressing my best thanks to Professor Williamson, under whose masterly guidance these experiments have been performed."[70] First, in scientific terms, Williamson most probably gave Sakurai guidance on possible research problems to tackle, and he definitely advised Sakurai on the choice of starter reagent from which a new substance would be prepared.[71] Sakurai's paper was concerned with the synthesis of organomercury compounds of a new class, that is, with a bivalent methylene radical such as $I(CH_2)HgI$. This topic fits with Williamson's research interests in molecular-structural problems, such as type and valency theories, and organometallic chemistry.

Second, Williamson gave several kinds of moral support to Sakurai. On August 20, 1880, Williamson wrote to Sakurai that "I am glad to hear that...you intend going to Swansea. I shall probably go down there on Tuesday. I have got the copy of your paper which you left for me and will give it to you when we meet there."[72] Here Williamson was encouraging about Sakurai's giving a presentation at the BAAS meeting in Swansea, an important debut platform for a young chemist. Williamson probably intended to give Sakurai comments on his paper and certainly informed Sakurai of common practices of BAAS contributors, such as submitting abstracts to presidents of sessions. Support of this kind continued after Sakurai returned to Japan in 1881 to take up an appointment at Tokyo University. According to a letter from Williamson to Sakurai dated September 19, 1882, Williamson read on Sakurai's behalf a further paper at the 1882 annual meeting of BAAS in Southampton and arranged its publication in the *Journal of the Chemical Society: Transactions.*[73] Undoubtedly he endeavored to give Sakurai the opportunity to develop and maintain his network of British scientists and to ensure promulgation of his research.

What Williamson apparently did *not* provide for Sakurai was coaching in experimental manipulation, day-to-day monitoring of his research, collaboration with him, or an opportunity for group research. Alan Rocke, in his reexamination of the Giessen model of university laboratory research, supplemented Jack Morrell's approach and explained that the success of Liebig's "chemist breeding" was due to the emphasis on group research; the latter was an essential element of Liebig's model of laboratory teaching.[74] Though he studied at Giessen with Liebig, Williamson did not seem to import this element of the Giessen model to UCL.[75] The design of the Birkbeck Laboratory at UCL, which Williamson inherited from George Fownes, another Giessen chemist, and where Sakurai received laboratory training, was itself congenial to group research and would have facilitated a professor's close monitoring of students' laboratory work.[76] Working benches were not partitioned for individual students, and the professor's private laboratory was contiguous with the main students' laboratory. However, according to Sakurai himself, "I do not remember seeing [Williamson] engaged in any experimental work while I was at University College.... and I believe I was almost the only research student working under him. I have, however, a faint idea that he was interesting himself with applying Dr. Barff's method of preventing iron from rusting (by treating iron with superheated steam and coating its surface with a thin film of magnetic oxide) to industrial purposes," presumably in Williamson's experimental works in Willesden in North London.[77]

This quotation, together with the close proximity of the students' and professor's laboratories, suggest that Sakurai rarely saw Williamson visit the Birkbeck Laboratory and that there was hardly any possibility for group research by Williamson and his students in his laboratory. For Williamson, the students' research work was basically their individual enterprise. In this sense, the chemical laboratory that was purpose-built for Williamson's teaching in 1881 arguably reflected the philosophy of his chemistry teaching more strongly.[78] The students' laboratory was separated physically from the professor's research laboratory and working benches were partitioned individually (hence the term "horseboxes" used by UCL chemistry students to refer to their benches). English architect Edward C. Robins commented that UCL's 1881 Laboratory "contains working-benches for fifty students, all of whom are as far as possible prevented from communicating with or overlooking one another; and in this respect it is quite unique."[79] Though he was not himself trained in this new laboratory, Sakurai's supervision of research students at Tokyo was to follow a similarly individualistic manner.

As the above quote shows, Sakurai was aware of Williamson's extramural industrial researches, and we have reason to infer that Sakurai was also interested in chemical technology and in the more general issue of the relevance of chemistry to the wider society. He was enrolled in the lectures on chemical technology delivered by Atkinson at Tokyo and by Charles Graham at UCL for two years, in 1878–1879 and 1879–1880.[80] Graham was to be one of the recommenders "from personal knowledge" of Sakurai when the latter applied in 1879 for election as a FCS.[81] Furthermore, on May 29, 1878, Sakurai was elected a member of the Society of Arts (Society for the Encouragement of Arts, Manufactures, and Commerce), which served as a platform for discussion of, among other things, the issues of technical education and the applicability of science to technology.[82] He remained a member until the 1879–1880 session.[83]

Judging from what Sakurai later wrote for the Japanese general public, however, Williamson's influence on Sakurai outweighed that of Atkinson or Graham. Sakurai seemed to support Graham's and Atkinson's argument that chemical technology could be an academic subject in its own right by using the expression "learning for practical uses" (*jitsuyō gaku*).[84] Yet Sakurai interpreted the relationship between this and "pure science" (*junsei rigaku*) just as Williamson did, i.e., technology as "the application of scientific principles to the complex processes that occur in manufacturing arts," and he described the relationship between science and technology as "there is no germination without seeds."[85] Sakurai never doubted that pure science was "seeds."

Based on this dichotomy between pure and applied science, Sakurai attached greater importance to pure science than to applied science in contrast to Atkinson and Graham. In his discussion of technological education (*kōgyō kyōiku*), just as Williamson had done in his *A Plea for Pure Science*,

Sakurai argued that, in general, engineers, who had little background in science, were unable to improve age-old manufacturing processes and would-be engineers had therefore to concentrate on scientific training during their university education before entering the business.[86] Though there is no direct evidence showing that Sakurai read Williamson's *A Plea for Pure Science*, the similarity between Williamson's and Sakurai's arguments is too striking to miss and reveals Sakurai's intellectual journey from Tokyo to London and his greater debt to Williamson.

Sakurai began to think about the social relevance of his subject during his student days in London. This suggests a different side of his training, namely, his interaction with wider English society outside classrooms and laboratories. He eagerly assimilated English culture, particularly through his personal acquaintances, and thereby underwent a process of socialization in English and more general Western cultural practices.[87] It was no wonder then that his Japanese students would regard Sakurai as a quintessential "English gentleman" (see chapters 5 and 6).

One of the gateways to socializing for college students is a student or alumni association. One such opportunity arose for Sakurai at an early stage. On November 9, 1876, just a few months after his enrolment in UCL, a preliminary meeting of students of the Birkbeck Laboratory was held under the chairmanship of Charles Graham to consider the formation of a society "for the reading of papers and for discussions upon Chemical matters," and this resulted in the establishment of UCL's Chemical and Physical Society.[88] Oliver Lodge, then a physics demonstrator, became its first president. This society was run by students and junior academics in both the chemistry and physics departments. They jointly organized reading seminars and series of addresses, which again testifies to the combination of chemistry and physics in UCL's science education. Sakurai was one of the founding members and gave an address entitled "Japanese Pigments" in the academic year 1879–80.[89]

More important, Masaki Taizō, superintendent of Japanese overseas students in London and Sakurai's former teacher in Tokyo, was also a founding member of this society. As I mentioned in chapter 1, during his own chemical studies at UCL between 1872 and 1874, Masaki had lived with Graham, who presided at the organizing meeting of this society. It is likely that Masaki would have taken the opportunity to do some active networking as a mediator by introducing Sakurai to other UCL students, alumni, and junior academics, including Graham, which again attests to Masaki's role as a "cultural mediator." Sakurai's election as a member of the Society of Arts in 1878 and as a FCS in 1879 meant a widening of the sphere of Sakurai's social as well as academic life in London.

Takamatsu at Owens College Manchester: Technological Chemistry in the Cottonopolis

After working for one year at the Tokyo Shihan Gakkō (Tokyo Normal School, a state school established by the Meiji government in 1872 for training teachers at the elementary level) as a chemistry teacher, Takamatsu was selected to be a government-sponsored overseas student of the Ministry of Education for three years, and he went to England in 1879. Takamatsu studied chemistry not at UCL, but in Roscoe's chemical laboratory at Owens College Manchester between 1879 and 1881 before he moved to the University of Berlin in Germany where he continued his overseas study until 1882.

At the college in Manchester, laboratory training during the period of Takamatsu's attendance followed a similar assistant-centered structure as in Williamson's laboratory at UCL, albeit in a more formalized and twisted way reflecting Roscoe's studies at UCL and Heidelberg. The two student laboratories of the Department of Chemistry at Manchester, designed in accordance with the arrangement prescribed by Roscoe, were located on the first floor and had "spy windows" connecting them to the professor's private room on the second floor that allowed the professor (director) to overlook and keep a close eye on his students' laboratory work.[90] Yet, it is doubtful that these windows played more than symbolic and psychological roles for the monitoring of students. An unpublished notebook of laboratory supervision made by Roscoe and his demonstrators (assistant lecturers) reveals the important role of the latter in laboratory monitoring.[91] Regardless of students' levels, day-to-day monitoring of their work was done by demonstrators, who assessed students' progress in knowledge, skill, and diligence and reported to the professor at the end of every session. Arguably, then, with the spy windows Roscoe wanted to give students the impression that they are being watched by him, but in reality he did not spend as much time with them as his mentor at Heidelberg, Robert Bunsen, had with him.

Regarding Takamatsu, this notebook indicates that he undertook original research right from the beginning. For the session 1879–1880, an unidentified demonstrator reported that Takamatsu was doing "research [on] pentathionic acid," measured "vap[our] densities" and "began org[anic] anal[ysis]." The writer added the comment: "rare perseverance, manipulative skill, and original power." Watson Smith, a newly appointed demonstrator, also wrote that Takamatsu was doing research and commented on Takamatsu's "rare perseverance & devotion" and "great analytical skill" in his report for 1880–1881.[92] Takamatsu clearly impressed his teachers with his analytical skill, which he had acquired at the Tokyo Kaisei Gakkō and Tokyo University.

One important implication of this system, which did not exist at UCL, was the encouragement of group research and the possibility of close collaboration between demonstrators and students, which was what Roscoe intended. A published history of the Department of Chemistry at Owens College written by Roscoe was regarded by Bud and Roberts as a clear sign that Roscoe wished to build a research school. It contained a catalogue of "communications made by demonstrators and students whilst at Owens College."[93] Roscoe undoubtedly considered the collaborative work by demonstrators and students an essential part of his "school."

Takamatsu coauthored four papers with Smith, and they were published in the *Journal of the Chemical Society: Transactions*; the papers covered scientific topics, such as the existence and constitution of pentathionic acid and phenylnaphtalene.[94] This choice of topics strongly reflected Roscoe's belief in the primacy of "pure science" and the value of research activity guided by this principle, as is shown by Takamatsu and Smith's acknowledgment of Roscoe in one of their papers.[95] This orientation during his Manchester years is also clear from Takamatsu's application for a FCS in January 1881, in which he identified himself as a "student of Scientific Chemistry," a term that had been commonly used to distinguish "purely scientific" chemistry from applied chemistry in Britain since the 1840s.[96] Takamatsu's close relationship with Smith, which later developed into a personal friendship, was based not only on their research partnership, but also on their mutual interest in technological chemistry.[97] It is therefore hardly surprising that Takamatsu took the course entitled "Technological Chemistry" delivered by Smith; he obtained the first prize in this course in 1881.[98]

The course called technological chemistry had as long a history as the Department of Chemistry and the college. Owens College Manchester was established in 1851 with the double aim of delivering general education along the line of English traditional universities and meeting local industrial demands. The first two professors of chemistry, Edward Frankland and Roscoe, basically met this challenge by adopting the liberal science model as a generic training course for a wide variety of chemistry-related professions under the heading of "systematic chemistry," "organic chemistry," and "chemical philosophy."[99] As both Frankland and Roscoe had substantial experience in consultancy and had strong connections to local chemical industries in Manchester, they tried to satisfy their local audience by delivering a course on technological chemistry, which taught the scientific principles underlying industrial processes in the alkali industry, dyeing, and dyestuffs manufacture, etc.

Based on this platform, however, Watson Smith introduced a new approach to technological chemistry in 1879.[100] After studying with Roscoe at Manchester and later with Georg Lunge, an expert in the alkali industry at the Zurich Polytechnic in Switzerland, Smith began his career at several chemical works in Britain while doing analytical researches mainly related to the coal industry.[101] Smith's course was totally different from those of his predecessors and also from the contemporary course on chemical technology of Charles Graham at UCL. Whereas Graham's course reflected a consulting chemist's view of technological education in chemistry, Smith's course reflected a works chemist's view of the same subject.

Smith's lectures did not deal solely with "applied" chemistry as the application of chemical principles and analytical technique to industrial processes. Instead, drawing on his previous experiences in both the alkali and coal-tar industries, he devoted a large part of his lectures to day-to-day plant operations at chemical works and the engineering aspects of the chemical industry. He dealt with the construction and description of chemical plants and apparatus and the planning and running of chemical works under the headings of "Alkali and Sulphuric Acid Manufacture; Bleaching Powder and Liquor; Potassium Chlorate; Carbon Bisulphide" and "Destructive Distillation of Coal; Gas Manufacture; Distillation of Coal-tar; Ammonia and Ammonium Salts from Gas-liquor." These headings suggests that, apart from the alkali industry, Smith seems to have been teaching about the "heavy" side of the organic chemicals industry, rather than the finer side, such as dyestuffs. Smith frequently organized industrial tours to chemical works, and he required students to submit drawings in his examination papers, which was not uncommon in civil engineering or mining subjects at that time, but unheard-of in chemistry-related courses.[102] The impact of Watson Smith's technological chemistry course on Takamatsu's career was to be apparent soon after he returned to Japan (see chapter 4).

There is another aspect that has to be mentioned regarding Takamatsu's overseas study. After two years at Manchester, Takamatsu moved to Germany and continued his studies at the University of Berlin for another year before returning to Japan. In Berlin he took experimental courses in organic and inorganic chemistry from August Wilhelm Hofmann, professor of chemistry at Berlin and former professor at the RCC, and Takamatsu also worked in Hofmann's laboratory. His application for a transfer to Germany to the head of the Ministry of Education, dated December 1880, reads:

> I have thus far studied chemistry at Manchester University and finished all the subjects I wished. So it would not be sufficiently beneficial for my study to stay at the same place. In contrast, in Germany chemistry is the most progressed subject of all and is well-organized particularly at universities. Therefore, I would like to go to this country and study advanced chemistry there.... I [have] discussed my future study plan with Professor Roscoe and others, who told me that moving to Germany was definitely the best way forward for my studies. [103]

Roscoe's advice to Takamatsu is predictable because Roscoe himself had done exactly the same thing following undergraduate study at UCL, and he believed that "scientific instruction in Germany [excelled] that of all other countries."[104] that Roscoe picked the German chemist Carl Shorlemmer as the professor of organic chemistry of his department. Owens College Manchester had a particularly strong German flavor among British colleges.

What Takamatsu did in Germany was therefore a natural extension of the training he received at Manchester with Roscoe and Smith. Takamatsu's biography states that he did experiments and research on the manufacture (*seizō*) of several dyestuffs and other organic chemicals in Hofmann's laboratory.[105] However, what he really investigated there was almost certainly not industrial methods of production on a mass scale, but the preparation of organic compounds on a laboratory scale.[106] Despite his close relationship with the dyeing industry through his former students, Hofmann at the

University of Berlin identified himself strongly as an academic chemist and saw "no reason to treat my academic subject other than in a purely academic manner."[107] As Jeffrey Johnson has argued, chemistry students in Hofmann's laboratory were trained exclusively in chemical analysis and organic preparations, leading his former students who pursued careers in the dyeing industry initially to model their industrial processes closely after the laboratory equipment and methods they had experienced in Hofmann's laboratory.[108]

However, Germany as a fast rising power in the chemical industry was definitely on Takamatsu's mind. Just before Takamatsu left England for Germany in July 1881, the first general meeting of the Society of Chemical Industry in London had been held in June. He was right in the milieu of the founding members of this society, as Roscoe was elected its first president, and Watson Smith was appointed the editor of its organ, the *Journal of the Society of Chemical Industry*, in 1882. Takamatsu became a founding member. In the presidential address, Roscoe expressed his academic view of the chemical industry as the application of the principles of pure science to technological problems and insisted that chemical manufacturers should do more of it. To make his case, Roscoe took the alarmist position that the chemical industry in England was losing ground to its German counterpart in branches such as the coal tar and synthetic dye industries.[109] That Takamatsu was impressed by this address is clear from a similar address he gave to the Tokyo Chemical Society in 1883, which I discuss in detail in the next chapter.

Takamatsu's decision to move to Germany was made before the establishment of the Society of Chemical Industry, but this society evolved out of earlier organizations in the northwest of England, including Manchester.[110] The above rhetoric of German supremacy in some branches of the chemical industry was common there around 1880. Takamatsu's turn to Germany would not be understandable without considering the milieu where Takamatsu was trained and from where one of the initiatives for this society emerged. The same penchant for Germany somehow echoed across the Atlantic in the overseas study of Takamatsu's fellow student, Kuhara at the JHU in Baltimore, Maryland.

Kuhara at the JHU: Organic Chemistry and the Americanized German Model

Though he was one of the five overseas students in the same year as Takamatsu, Kuhara had reasons to consider his study as something special. His overseas studies were very likely solicited and partly funded by a host university in the United States. According to his own explanation to his father, "the President of the university in Baltimore, Maryland of America [i.e., JHU] recently wrote to the Ministry of Education that, if the ministry would select and dispatch the most academically accomplished student from the graduates of Tokyo University, they would educate this student to be a great scholar. And I was selected specifically for this position."[111] He continued with apparent excitement that the payment (not a loan as was usual in this period) of his study costs would be made jointly by the Japanese Ministry of Education and the JHU. If it was true, it surely was exceptional, if not impossible, under the conditions of the time.

Though I could not find corresponding sources in the archive of JHU's first president, Daniel Coit Gilman, I find it more than likely that he wrote such a letter to a foreign government.[112] First, in 1879 JHU was quite a young university and had been in operation for only three years and therefore had a small student body. This kind of action from a president was actually not as unusual as it may sound.[113] Perhaps more important, a steady supply of well-trained graduate students was essential for the growth of the research school of its first chemistry professor, Göttingen-trained Ira Remsen. JHU's fellowship program, which modestly benefited Kuhara and other fellows with $500 each per annum, was just one of many measures to ensure such a supply.[114]

In this respect, it is worth noting Kuhara's strong conviction that his alma mater, Tokyo University, was on a par with American colleges. This conviction was expressed in his letters to his parents after visiting Columbia College School of Mines in September 1879. He went so far as to write: "Before departure I had imagined how advanced American schools would be and what great scholars would inhabit them. Contrary to what I had imagined, there is no particular difference between them and Tokyo University. Rather, in some points American schools were less advanced than Tokyo University. This shows that Japanese universities are now on a par with American ones, so it is no longer of much benefit to dispatch Japanese overseas students to the United States."[115] Putting his self-congratulation aside, his confidence in competing with American graduates as a laboratory worker at the JHU is clear. Indeed, Kuhara got straight into laboratory work as a fellow after he entered the JHU in fall 1879 and only took a course in analytical chemistry offered by Harmon N. Morse, another Göttingen-trained American chemist who worked as Remsen's associate.[116] Kuhara was exactly the kind of "workforce" Remsen would have needed.

Remsen has been rightly recognized as the founder of American graduate education in chemistry after the German model, and historians added further subtlety to this image. Owen Hannaway argued that what Remsen established at JHU was not a replica of a German university but built on "an ideal refracted through a prior American experience," that is, an Americanized German model. He perceptively observed that this was because "what in the end drove Remsen's career and shaped his perception of it, was his own overwhelming need to define him a professional man in a country with an underdeveloped sense of a professional ethos."[117] Deborah Warner developed this theme by arguing that while Remsen carefully presented the image of his "disinterestedness" to distinguish the JHU from other practice-oriented schools such as Columbia's School of Mines, he was "not disdainful of the many improvements that the 'art' of chemistry brought to daily life."[118] On the contrary, he aggressively claimed credit for important practical results, in his case the invention of saccharin, against his erstwhile coworker Constantin Fahlberg by conveniently ignoring the enormous problems of turning a laboratory finding into a commercial commodity. It is symptomatic that Remsen discussed possible lawsuits for this purpose with Chandler, one of the expert witnesses in industrial chemistry as seen above.[119]

Two characteristics of Remsen's pedagogical approaches are important, both carried over from Remsen's experiences in Göttingen. First, to keep students abreast of the latest literature, Remsen held "reading and discussion of current chemical journals" twice a week throughout the year for 1879–80, which was later renamed "Journal Meetings."[120] In his letters Kuhara repeatedly told his parents how much he benefited from discussion with his peers and from reading chemical literature in German and French in addition to English.[121] He wrote about social events held every Saturday evening and attended by teachers and advanced students, including Russians, Britons, and Germans. Kuhara said with apparent joy that they drank beer, talked about their own countries, and sang songs. He added that "every country is the same and no different from Japan," a recurrent theme in his letters to his parents.[122]

Once Kuhara entered the laboratory at JHU, however, things were quite different from Tokyo. As Warner pointed out, Remsen made the laboratory a "hierarchically structured social space" by laying out "factory-like" regulations. In addition, to regulate students' progress in the laboratory, he assigned projects and gave instructions during his daily visit to the desk of each research student.[123] Kuhara, in his letter to his parents, wrote that "My teacher, Mr. Remsen, ... once studied in Germany and graduated from a university there. Therefore, he is one of the most brilliant scholars in America. In addition, he is very kind and attentive to students in his instruction."[124] In pedagogical situations "kind" "attentiveness" and tight control are often two sides of the same coin.

The biggest difference between Atkinson's and Remsen's laboratories was that Remsen, as the laboratory director, often assumed lead authorship in students' publications.[125] The question is who "possessed" which research projects done in a laboratory. The analysis of Kuhara's four papers produced while he was at JHU and published in *American Chemical Journal* gives us a glimpse of exactly how this worked. His first paper, "A Method for Estimating Bismuth Volumetrically," was published in December 1879.[126] The publication date and analytical nature of this paper makes it more than likely that this project came out of his analytical training with Morse.[127] Nevertheless, Morse was not mentioned in the article, and it was published with only Kuhara's name.

The remaining three papers, "Sulphoterephthalic Acid from Paraxylenesulphonic acid," "Concerning Phtalimide," and "On the Oxidation of Substitution Products of Aromatic Hydrocarbons. XL.: On the Conduct of Nitro-Meta-Xylene Toward Oxidizing Agents," were published in February 1881, March 1881, and February 1882, respectively, and the last one of them was a modified form of Kuhara's dissertation.[128] To varying degrees all three originated in Remsen's own research program on whether a base in a benzene compound had a "protective influence" upon another oxidizable base in the ortho or meta position when subjected to the action of oxidizing agents such as potassium permanganate.[129] Therefore, it makes sense that the second and fourth papers named Remsen as the lead author. The question, rather, is why the third paper, "Concerning Phtalimide," was published with Kuhara's name only.

The clue to the answer is the distance between paper topics and the mainstream of Remsen's research program. Kuhara's dissertation, "On the Conduct of Nitro-Meta-Xylene Toward Oxidizing Agents," addressed exactly the same question as Remsen's and "confirmed" the protective influence of nitro, amino, and hydroxyl bases as well as chlorine and bromine upon a methyl base in the ortho position. This also means that Kuhara likely received advice from Remsen frequently. His second paper, "Sulphoterephthalic Acid from Paraxylene-sulphonic Acid," was written essentially to substantiate the claim of Remsen and W. Burney that they prepared sulphoterephthalic acid (which does not contain nitrogen) by oxidizing sulphoamineparatoluic acid (which contains nitrogen) with potassium permanganate. Kuhara did so by preparing the identical substance without using compounds with nitrogen.[130] Though his paper exhibits his experimental ingenuity, it is so subordinated to Remsen's and Burney's article that it is impossible to understand the object of the former without reading the latter.

"Concerning Phtalimide" is different. Kuhara acknowledged Remsen's "suggestion," but this project really started from Remsen's and Fahlberg's celebrated paper announcing the preparation of benzoic sulfinide or saccharin, the project in which Fahlberg took the lead.[131] Apart from the method of the substance's preparation by oxidizing orthololuenesulphamide, this paper mentioned another way of getting benzoic sulphinide, namely, by treating orthosulfobenzoic acid with phosphorus pentachloride and ammonia, not by oxidization.[132] Using the same procedure, Kuhara started with phtalic acid instead of orthosulfobenzoic acid and obtained an unknown isomer of phthalimide.[133] Kuhara's third paper therefore represented a sideline of Remsen's research program on the oxidation of aromatic compounds. That is presumably why Kuhara could publish the paper in his own name.

Professors' lead authorship in students' works at the JHU chemical laboratory therefore depended neither on how much students contributed to devising experimental procedures nor on teachers' support of students' experiments. It depended on how closely a particular research project was connected to the research program of the laboratory director, who "owned" the program. Students would get into this research program and work on a (tiny) part of the whole to the point that they could not understand on their own what their work was supposed to achieve. That is what Kuhara learned there.

How did Kuhara react to this "hierarchically structured social space"? Both in printed form and in private correspondence, he did not show any sign of complaint or resentment. Rather, he sang the praises of JHU's "advanced [*kōjō*]" and "carefully structured [*yukitodokitaru*]" teaching regulation as compared to that of other American colleges like Yale College.[134] He most likely internalized Remsen's pedagogical regime and used it in constructing his research school at Tokyo University and Kyoto Imperial University.

Despite the fact that Matsui, Sakurai, Takamatsu, and Kuhara all attended the same school at Tokyo, differences in their experiences were such that, after they returned to Japan as professors of chemistry at Tokyo University, they would play totally different roles in the institutionalization of scientific and technological education in chemistry.

Chapter 4

Defining Scientific and Technological Education in Chemistry in Japan, 1880–1886

When Matsui Naokichi, Sakurai Jōji, Takamatsu Toyokichi, and Kuhara Mitsuru returned to Japan to take up a lectureship in the Faculty of Science of Tokyo University in the early 1880s, the government-controlled Japanese higher education system was in the early phase of a restructuring process. This was no coincidence. A wide-ranging deflationary policy initiated during the 1880s by Finance Minister Matsukata Masayoshi included government retrenchment and the introduction of a cabinet system of government in 1885. In this context, the Meiji government and its schools replaced the more costly Western professoriate with Japanese teachers and unified the educational system hitherto dispersed under various ministries in order to eliminate waste of teaching resources.[1] This rationalization culminated in the merger of Tokyo University run by the Ministry of Education with the Imperial College of Engineering, Tokyo established by the Ministry of Public Works into the Imperial University in March 1886.

The timing of these events provided the British- and American-trained Japanese chemists with both opportunities and challenges. It enabled them to participate in the process at a crucial point and to implement their visions of chemical education nurtured during their training in Japan, Britain, and the United States. However, they also had to respond to the ever changing institutional situation outside of their control. This was the period of chemists' soul-searching in the crucible of history.

Two important events of institutional restructuring for the Faculty of Science occurred before the establishment of the Imperial University in 1886. The first was the geographical transfer of the faculty in the summer of 1885 from the Kanda Nishikichō campus to the Hongō campus so as to share the site and buildings of the Faculty of Medicine. The main effect of this move on science professors was to bring them into contact with medical professors as colleagues; among those professors was the German-trained pharmaceutical chemist Nagai Nagayoshi. Nagai became involved in the planning of a new general chemical laboratory that was to be shared by all the chemistry-related departments of Tokyo University. However, the plan for collaboration among these departments proved short-lived; it was superseded by the establishment of the Imperial University as a federation of the subject-based colleges (*bunka daigaku*) of law, medicine, engineering, literature, and science. The laboratory that Nagai had been planning then became the main building of the College of Science and, through its design, would have an impact on how Sakurai would later structure his pedagogical space. I will tell that story in chapter 5.

The other event affecting the organization of the Faculty of Science in this period was the separating out from it in December 1885 of a new Faculty of Industrial Arts (*Kōgeigakubu*). This had long-term implications for the institutionalization of scientific and technological education in chemistry, as the creation of a second faculty was closely related to the key question of how best to organize higher-level scientific and technological education, the main focus of this chapter. In the event, it was this new faculty that was merged with the Imperial College of Engineering, Tokyo in March 1886 to form the College of Engineering of the new Imperial University.

In order to analyze the development of teaching programs in chemistry between 1880 and 1886 in the wider institutional context, I shall consider two issues here: (1) the different ways in which British- and American-trained Japanese chemists translated their visions of chemical education into institutional models, and (2) the different roles played by expert teachers and government ministers and officials in defining scientific and technological higher education in chemistry. The most important actors on the administrative side were Mori Arinori and Watanabe Hiromoto (or Kōki). Mori was a former Satsuma student in Britain and the United States and the first education minister in Itō Hirobumi's cabinet. Watanabe was handpicked by Mori as the first president of the Imperial University.

Appointing the First Japanese Professors of Chemistry

All four Japanese professorial appointments at Tokyo University came a year after these scientists had been appointed as lecturers in 1880 (Matsui), 1881 (Sakurai), and 1882 (Kuhara and Takamatsu).[2] Actual teaching began with their lectureship, but promotion signaled that they had assumed full responsibility and freedom in offering course(s) in the Department of Chemistry; therefore, it is important to note both years to establish a chronological framework.

It is equally important to note that, though both Matsui's and Sakurai's appointments were considered by university officials as replacing Jewett and Atkinson, Sakurai succeeded Atkinson only in his role as professor of analytical chemistry (*bunseki kagaku*) when appointed lecturer in 1881. Atkinson's professorship of applied chemistry (*ōyō kagaku*) was made independent, and it was renamed professorship of chemical technology or "manufacturing chemistry" (*seizō kagaku*) and given in 1881 to German industrial chemist Gottfried Wagener. It was this post that officially went to Takamatsu in 1884, when Wagener moved to a professorship of chemical technology at the Tokyo Shokkō Gakkō, a secondary-level technical school established in 1881 by the Ministry of Education. But Takamatsu had already been involved in the teaching of this subject since 1883, the year he was promoted to full professor. Kuhara was also appointed full professor of chemistry at Tokyo in 1883, and he then started to take a substantial role in departmental teaching.

A Ministry of Education document dated February 1881 provides insight into its policy for the selection of professors of chemistry in the early 1880s and into the qualities required of a teacher of this particular subject:

> As for substitutes for Atkinson, we have a prospective candidate for professor of analytical chemistry [i.e., Sakurai] among overseas students who are due to return to Japan soon. But we have no Japanese candidate for professor of applied chemistry (that is, chemical technology). So we propose to hire a German, Wagener, as professor of this subject from the First Day of April for two years with the salary of 430 *yen* per month in paper money. . . . This person is familiar with chemical technology, especially the practice of manufacturing industry in this country. So we propose to hire this person as soon as possible to fill the gap.[3]

A number of points emerge from this quotation. First, by this time Westerners were appointed only if there were no suitable Japanese candidates. Second, both the University and the Ministry were interested in teachers of applied chemistry well-versed not only in chemistry itself but also in the practices of manufacturing industries in Japan. Ironically, they could only find such a qualified teacher among foreigners but not among their pool of Japanese students.

Lastly, at this time both Tokyo University and the Ministry of Education viewed the Japanese professors as substitutes for Westerners, rather than expecting them to add any new dimensions to chemical education. Chemico-analytical training at the Tokyo Kaisei Gakkō and Tokyo University was strongly oriented toward practical uses, either for commercial, industrial, or public purposes (chapter 2). The above quotation emphasizes the professorship of applied chemistry and does not indicate that education officials wanted any change in analytical training. Nevertheless, it is a potentially significant step to separate professorships of analytical and applied chemistry because this provided Sakurai with the opportunity to reorient his teaching in the direction he preferred, asserting the value of pure chemistry. This would have been difficult, if not impossible, if he had been given a double appointment.

From the candidate's perspective, the selection process would not seem to have been straightforward. In his autobiography, Sakurai recalled receiving the job offer, probably in 1881, during his overseas study in London from

Katō Hiroyuki, who was president (*sōri*) of Tokyo University from 1881 until 1886:

> I received a letter from Katō Hiroyuki sensei [common courtesy title in Japanese for teachers or seniors], President of Tokyo University. It told me that he wanted me to teach at Tokyo University. . . . I thought that I, a young man of only 23 years of age with only limited knowledge and ability, was mistakenly invited by the university president and very much hesitated to accept the offer. I eventually accepted this offer just because I did not have the courage to decline it.[4]

Sakurai's words have a strong ring of humility (*kenson* or *kenjō* in Japanese), which was interpreted as the result of his being "highly deferential to authority."[5] Was he really chosen by chance or was some selection process at work?

No university or government records show the process of Sakurai's and other professors' selection. However, the Tokyo Kaisei Gakkō and Tokyo University did keep fairly detailed academic records of overseas students. These were received from the superintendents of the government-sponsored overseas students dispatched between 1875 and 1879 by the Ministry of Education.[6] Some of them even went to the *Dajōkan*, the central executive office of the Meiji government. The records covered the courses the students took, their examination grades, rankings, prizes, and degrees as well as information about any original research projects (*shinkō shiken*) they undertook.

Among government overseas students in chemistry in this period, Sakurai appears in these records as having good credentials as a candidate for professor of analytical chemistry. In addition to achieving a prize, stipend, and a membership in a specialist society, Sakurai worked as what we would now call a teaching assistant in Williamson's class of analytical chemistry at his mentor's request.[7] The superintendent of Japanese overseas students in Britain, Masaki Taizō, seemed particularly impressed by Sakurai's receipt of the Clothworkers' Exhibition award in 1879, and he sent a special report to the ministry separately from his routine reports.[8] Though Sakurai could not enroll for a University of London degree, the Ministry of Education seems to have regarded the successful completion of his chemical studies as equivalent to graduation.[9] Matsui and Kuhara both finished their studies with a Ph.D. and thus were a more obvious choice. The first prize in Watson Smith's class of Technological Chemistry in 1880 may well have appeared to be just the right credential for Takamatsu's appointment as professor of chemical technology.[10] None of the candidates seems to have been "mistakenly" selected to a post as one of the first Japanese chemistry professors for Tokyo University.

A Study in Contrast: Matsui's and Sakurai's Participation in the Teaching Reform at Tokyo

Matsui was appointed lecturer at Tokyo University in 1880 with the unenviable job of taking over much of the departmental teaching from Jewett,

Atkinson, and their Japanese assistants. This task included delivering lectures on inorganic chemistry and supervising chemical experiments for all science students as well as lectures on organic chemistry for chemistry students, and supervising laboratory sessions in qualitative analysis for students of chemistry, mining, geology, and physics. The last duty meant that Matsui had to adapt his training to different levels of students, not an easy task for a novice teacher.[11]

Matsui took an incremental approach to improve his teaching environment by asking Tokyo University for permission to increase contact hours with students and to use a better-equipped chemical laboratory (with coal gas instead of charcoal in the university laboratory) at the Geological Survey of the Ministry of Agriculture and Commerce.[12] His skill and innovation in blackboard teaching is exemplified by his clear-cut lecture in Japanese on the complicated topic of chemical atomism in the first half of the nineteenth century, a lecture he gave at the 1882 annual meeting of the Tokyo Chemical Society (*Tokyo Kagakukai*).[13] This was a specialist society for chemistry founded in 1878 with Kuhara (before moving to JHU) as its first president.[14] It was this lecture that presented Mendeleev's periodic law in print for the first time in Japan.[15]

In a modest way Matsui put his stamp on university teaching using his experiences at Columbia and Yale; for example, he offered a one-off new course on physiological chemistry in 1880–81 to biology students that focused on the physiology of animal and vegetable fluids and tissues, and he assigned a research project of analyzing drinking water in Tokyo to chemistry students in their fourth, and last, year in 1882–83.[16] However, perhaps partly because of overwhelming teaching duties at the beginning of his career, Matsui was not keen on adding extra courses to his offerings and showed little interest in radically changing the curriculum of the faculty and department. Conspicuously missing is his participation, both inside and outside of university, in the discussions of the best teaching methods for applied chemistry and chemical technology, the very subject he majored in when studying in the United States.

In contrast, once Sakurai took up his position in Tokyo, he did not hesitate to express his views on chemical education. In December 1881, while still a lecturer teaching inorganic chemistry and qualitative analysis to junior science students, he read a methodological paper entitled "Discussing the Relationship between Chemistry and Physics" at an 1881 meeting of the Tokyo Chemical Society under Matsui's presidency.[17] The paper was published in the *Tōyō gakugei zasshi*, a periodical for popularizing science in Japanese, whose editors and contributors were Tokyo University academics. Thus, this presentation was originally prepared for audiences consisting of Sakurai's fellow Japanese chemists and later published for the general public. The paper is important because it reveals Sakurai's outlook on chemical education shortly after his period of overseas study, and it merits detailed analysis here.

Sakurai felt that a gradual recognition among the Japanese general public of the educational value of science teaching made his discussion of the relationship between chemistry and physics pertinent:

> The public is gradually admitting that studying science [*rigaku*] is urgently needed in terms of education as well as practical use. For example, nowadays introductory lectures in science are given not only at universities and middle and normal schools, but also at elementary schools in our country. In teaching science, it is essential to use an appropriate method.[18]

According to him, the "appropriate method" was to begin with the most general subjects and to proceed to more specific subjects. He then introduced his main argument—that one could not master chemistry without studying physics and mathematics.

Sakurai admitted that certain aspects of chemistry, i.e., the art of chemical analysis, could be mastered without the assistance of physics. However, he argued that "chemical analysis alone does not constitute the aim of chemistry, because chemistry studies the changes caused by the vibration and motion of atoms."[19] Sakurai explained that though in the past chemists had taken the lead in revolutionizing the understanding of atomism by introducing Daltonian atoms, which were distinctly different from Greek atoms, in the future chemists would have to clarify the truths of chemistry from the viewpoint of physics to keep their lead and advance the science further. To corroborate this point, he gave the example of the diatomicity of common molecules, such as hydrogen, oxygen, and chlorine, which had been introduced by chemists, but then confirmed by physicists using the kinetic theory of gases. To conclude, Sakurai insisted on the centrality of physics and chemistry to science as a whole.

Sakurai's arguments here seem unsurprising in light of his familiarity with Williamson's work. First, in formulating his definition of chemistry, Sakurai borrowed almost verbatim from Williamson's dynamic view of the atomic constituents of molecules, derived from his famous study on etherification.[20] Second, Williamson shared with Sakurai a keen interest in teaching methods and had strong views on the subject; Williamson's approach might be labeled "the inductive method," as I discussed in chapter 3. Third, the combination of chemistry with physics and mathematics in teaching was also a key component of Williamson's view of chemical education.

Lastly, as Sakurai observed that "the public [in Japan] is gradually admitting that studying science is urgently needed in terms of education," science teaching would become an important career destination for Sakurai's students. This was consistent with Williamson's view of the vocational potential of university science education. Not long afterward, in 1884, Sakurai gave a further lecture entitled "How to Teach Chemistry," which used Williamson's "inductive" method, at the Education Society of Great Japan (*Dai Nihon Kyōikukai*), the first association of teachers at elementary and secondary schools in Japan.[21] In this lecture, Sakurai argued:

> What chemistry teachers have to do most is not to explain the principles of chemistry first, but to show these principles through practical demonstrations and experiments. And the principles these experiments are to show should be as simple as possible so that pupils' minds can accept them easily and understand their meaning. Chemistry teachers do not teach these principles to pupils. Pupils learn them, and teachers are but a guide for pupils' learning.[22]

He thus started to form a social network within the educational sector through his activities to promote Williamson's ideas on chemistry teaching.

However, Sakurai's theoretical and physicalist approach to chemical education was novel in a different context, i.e., the regime of chemical education in Tokyo. The curriculum of the Department of Chemistry at Tokyo University in 1881–1882, when Sakurai was appointed lecturer (see table 4.1), centered on chemical analysis and chemical technology and did not provide any room for Sakurai's ideas, except a brief mention of the relationship between physical and chemical properties of inorganic compounds in his lectures on inorganic chemistry.[23] Sakurai's assertion above that "chemical analysis alone does not constitute the aim of chemistry" may well suggest some dissatisfaction with his position as lecturer of analytical chemistry to junior students. Indeed, as soon as Sakurai was promoted to full professor in 1882, he introduced a new lecture course on Chemical Philosophy for advanced chemistry students, which lasted until 1885–1886 and was a first step toward the implementation of his vision.

Sakurai's new course would start with the history of chemistry and contemporary chemistry, thus building partly on the nineteenth-century European tradition of introducing the study of chemistry by means of historical discussion.[24] However, the course would soon turn to the section on "contemporary chemistry," which covered several theories of recent "modern" chemistry. Among these were valency theory, the periodic law of Mendeleev (more or less at the same period as Matsui presented it in publications), the relationship between chemical and physical properties of substances (to which Sakurai promised to pay particular attention), and thermochemistry.[25] He stressed that these lectures were assembled from original papers and abstracts rather than from textbooks.

Lecture notes of this course, taken meticulously by one of Sakurai's students between 1882 and 1883, show important details.[26] For example, just like his teacher Atkinson, Sakurai used both type and graphic formulas (with bonds) in his explanation of valency (or "atomicity," as Sakurai termed it) and organic structural theory. The presentation of theories was understandably centered on Williamson's water type and his famous paper on etherification. As a new feature of his course, the measurement of molecular weight by physical properties and the confirmation of the diatomicity of common molecules by the kinetic theory (which he had mentioned in his 1882 article) were explained in detail. Indeed, understanding his explanations would have required some theoretical knowledge of physics. The relationship between

Table 4.1 Curriculum of the Department of Chemistry, Tokyo University, 1881–1882

	First Year	Second	Third	Fourth
Mathematics (Analytical Geom.)	Kikuchi/Miwa			
Elementary Mechanics	Ōmori			
Astronomy (Outline)	Paul			
Physics			Yamagawa	Yamagawa
Elementary Mineralogy	Wada			
Elementary Geology	Gottsche			
Mineralogy		Brauns/ Gottsche		
Metallurgy			Netto	
Drawing	Taga			
Logic	Chigashira			
English	Toyama/ Houghton	Houghton		
French			Koga	
German		Iwasa/ Senn	Iwasa/Senn	
Inorganic Chem. (Lectures + Labs)	Sakurai			
Organic Chem.			Matsui	
Chem. Technology			Wagener	Wagener
Analytical Chem. (Qualitative)		Sakurai		
Analytical Chem. (Quantitative)		Matsui	Matsui	
Analytical Chem. (Assaying)				Iwasa
Blowpipe Analysis		Iwasa		
Graduating Thesis				○

Sources: TDn, vol. 2; *Jpn. & Eng. Calendars Tokyo Univ. Law, Sci., and Lit.*

physical properties (such as specific volumes, melting and boiling points, and optical properties) and chemical constitution was particularly well covered, as was thermochemistry. The coverage of this latter topic was later extended in the final presentation of this course in 1885–1886.[27]

The relevance of this lecture course to Sakurai and to Tokyo's Department of Chemistry in later years is twofold. First, as I discuss in chapter 6, after the establishment of the Imperial University, Sakurai would develop this course into a course on theoretical and physical chemistry by incorporating ideas and methods from Ostwald's new school of physical chemistry.[28] Second, Sakurai "spun off" popular lectures and articles from his academic lectures on chemical philosophy for the purpose of presenting his views to his fellow chemists as well as to the general public. The common theme of these lectures was his physicalist and dynamic definition of chemistry. For example, his presentation in English titled "On Thermochemistry" at the 1883 Annual Meeting of the Tokyo Chemical Society was translated by his students and published in the society's organ, the *Tokyo kagaku kaishi* as well as in *Tōyō gakugei zasshi*. These presentations from the section of his course devoted to thermochemistry concluded with an explicit citation of Williamson, whom Sakurai referred to as "a chemist enthusiastic about thermochemistry," that all chemical changes should be interpreted in terms of atomic movements.[29]

Sakurai went further in his presidential address in Japanese at the April 1885 Annual Meeting of the Tokyo Chemical Society. After reviewing the latest developments in chemistry with materials from the "contemporary" part of his course, he expressed his opinion about the future prospects of chemistry and introduced the changes he was implementing in the curriculum of the Department of Chemistry at Tokyo University, explaining his physicalist approach to his fellow chemists in Japan:

> This review shows that, however astonishing they might be, the recent developments in chemistry reviewed here all belong to chemical statics [*kagaku seishigaku*] and that chemical dynamics [*kagaku undōgaku*] still remains an uncultivated area. We cannot expect advancement of chemistry in general without progress in the latter branch of chemistry.
>
> How can we make progress in chemical dynamics? The only way forward is to study chemical phenomena by using mathematics and physics. According to my opinion, everyone engaged in pure chemistry [*junsei kagaku*] has to study these two subjects in depth. This is why the subjects of calculus and advanced physics were introduced in the curriculum of the Department of Chemistry at Tokyo University a few years ago. [30]

This is Sakurai's first public use of the term *junsei kagaku* (pure chemistry). As I discuss later in this chapter, a separation of pure from applied chemistry had been introduced in the Department of Chemistry at Tokyo University in August 1883. Here, Sakurai was reporting it to fellow chemists.

Sakurai's assertiveness is clear from the above quotation. But he reserved his strongest message for the very end of his address:

> If chemists are content with analysis and synthesis and stop their research [in chemical dynamics], chemistry will become a dead science, and the study of atomic movement, the basis of chemistry, will be entirely in the hands of physicists.[31]

To sense the boldness of Sakurai's message, we only have to see that most of the contributions to the organ of the Tokyo Chemical Society, *Tokyo kagaku kaishi,* were concerned either with conventional analysis and synthesis or with chemical technology with analysis as its main component.[32] Until 1885 the contributors to the *Tokyo kagaku kaishi* were almost all alumni of the Department of Chemistry of Tokyo University. It is therefore likely that most of their publications derived from their own graduate theses, and we may infer that analysis and synthesis were the principal activities of students of Sakurai's own department.[33]

The image of Sakurai that emerges from the above discussion can hardly be described as "deferential" or "reconciliatory," as had once been suggested. The young Sakurai adopted a strategy of transfer without translation, or direct rendering into Japanese without modification, of Williamson's ideal of pure science at several levels and tried to transform chemical education in his department rapidly in a matter of years. For this purpose he was prepared to attack the traditional emphasis on analysis and synthesis in chemical research. A likely target of Sakurai might well have been none other than his colleague at the Department of Chemistry, Kuhara Mitsuru.

Kuhara's Vision of Chemical Education: Organic Chemistry and Journal Meetings

Shortly after Sakurai expressed the physicalist vision of chemistry and its teaching for the first time in December 1881, Kuhara gave a lecture in April 1882 at the same Tokyo Chemical Society in which he presented an alternative vision that somehow overlapped with Sakurai's but had different emphases and nuances. Aptly titled "Studies of Organic Chemistry," Kuhara's address argued the centrality of organic chemistry, not physics and mathematics, in chemistry.[34]

Starting with explaining organic chemistry's origin in chemical studies of substances extracted from animal and vegetable bodies, Kuhara first made the point that it had now become the chemistry of carbon compounds and was therefore no different from other parts of chemistry in its connection to the whole subject. In support of this argument he cited Roscoe and Schorlemmer's *The Chemistry of the Hydrocarbons and Their Derivatives.*[35] Organic compounds undergo the same kind of reactions and obey the same chemical laws as iron or sulfur compounds, to use his example.

Kuhara then got straight to the point that it is necessary for all chemists to study organic chemistry thoroughly to understand the overall principles of chemistry. According to him, recent theoretical breakthroughs in chemistry had been almost exclusively due to developments in organic chemistry that gave rise to such key concepts as substitution, isomerism, constitution, and valency.[36] He then gave a brief historical overview of organic chemistry in a similar manner as Sakurai was about to do in his lectures on chemical philosophy, from the type theory to the structural theory through valency. However, Kuhara then proceeded to make his own point that the study of organic chemistry would be central to theoretical chemistry.

One powerful example Kuhara used is the interlocking relationship between isomerism and constitution. Because isomers designate two or more compounds with the same empirical formula and same molecular weight but with different properties, he could argue that studying organic compounds to painstakingly gather information about their properties is indispensable for establishing isomerism and therefore constitution, the latter of which is a matter of theoretical chemistry.[37] As he put it, "one reason why chemists often hold mistaken opinions on the constitutions of compounds is that they do not study carbon compounds well enough."[38]

Thus far Kuhara was talking about his viewpoint developed throughout his training at the Tokyo Kaisei Gakkō, Tokyo University, and the JHU. Many in his audience would not have had any difficulty following his presentation. From there his speech took on a distinctively American flavor. To corroborate a corollary of the last point, i.e., that chemists should be well informed about the latest research on the properties of carbon compounds, he likened chemists who stopped reading journal articles to Rip van Winkle, the protagonist of the famous short story by American author Washington Irving who stayed in the mountains while the colonists fought the American Revolutionary War and thus lost contact with current affairs.[39] This was actually a humorous introduction to JHU's Journal Meetings (*Zasshi-kai*) that Kuhara found so beneficial while he was there (chapter 3). According to his straightforward explanation, those meetings were nothing more than gatherings to read through chemical journals. Each graduate member was assigned one journal each and had to present simple summaries in a meeting held twice or so a week.[40] The importance of Journal Meetings for the student life of Tokyo's Department of Chemistry from the 1890s on is discussed in detail in chapter 5.

The last, but no less important, point Kuhara made concerns the dichotomy between pure chemistry (*junsui kagaku*) and practical chemistry (*jitchi kagaku*) or "manufacturing chemistry" (*seizō kagaku*). As he was giving this lecture in Japanese, an awkward but more literal translation of the last term, "manufacturing chemistry," better conveys Kuhara's rhetoric than the term "chemical technology." Here, too, Kuhara exhibits traits of his training at the JHU: Essentially Kuhara saw no difference whatsoever between the teaching of pure and manufacturing chemistry and advocated the teaching of the latter as a subject of organic chemistry by drawing on the examples of the

German-language *technische Hochschulen*, such as Berlin and Karlsruhe.[41] In a way reminiscent of Remsen, he proclaimed near the end of his address:

> Manufacturing chemistry also progresses along with the developments in the chemistry of carbon compounds. Without doubt, with the progress in the chemistry of carbon compounds someday in the future we will be able to procure artificially the things we cannot make without plants today.[42]

Like his mentor, Kuhara here ignored the laborious process of commercializing synthetic products into commodities.

Kuhara's address can be interpreted on different levels. As an address given to his Japanese colleagues only a couple of months after he returned from the United States, it directly reflected his experiences at the JHU and the values he internalized there. On one level his address is a counterargument to Sakurai regarding which subdiscipline should be at the center of chemical education and research. On another level, the lecture is an argument of the supremacy of pure chemistry, much like Sakurai's lecture. On yet another level the address can be read as a plea against separating the teaching of pure and manufacturing chemistry based on the perceived supremacy of organic chemistry qua pure chemistry. It is this last issue that Kuhara's and Sakurai's colleague Takamatsu was to address a year later, in a different way.

Takamatsu's Vision of Education in Technological Chemistry

Takamatsu's address in question, given at the April 1883 Annual Meeting of the Tokyo Chemical Society while he was still a lecturer, was entitled "On the Method of Studying Manufacturing Chemistry [*seizō kagaku*]."[43] Again, it is better here to adopt "manufacturing chemistry" rather than "chemical technology" as the translation of *seizō kagaku* to understand Takamatsu's rhetorical problem: It would simply be easier to say that pure chemistry and chemical technology are different than to do the same for pure and manufacturing chemistry. Kuhara made the most of these rhetorical minutiae, but Takamatsu had to deal with them differently because his central message, which came at the very end of his address, was the plea for instituting a specialized program for *seizō kagaku* within the Department of Chemistry of Tokyo University.[44] His way of presenting his arguments to his fellow chemists is distinctively different from that of Sakurai and Kuhara.

Takamatsu's use of rhetoric requires some effort to decode it. He started with a definition of "manufacturing chemistry," stating that it was called "technical chemistry," "technological chemistry," or "applied chemistry" in English and therefore had the same meaning as *ōyō kagaku*, then the standard Japanese translation of "applied chemistry":

> The main object [of manufacturing chemistry] is to deal with the methods of producing useful goods from natural raw materials or waste by

> applying the principles of chemistry [*kagaku no ri*]. That is, manufacturing chemistry is a branch of pure chemistry [*junsei kagaku*] and therefore manufacturing chemistry progresses just as pure chemistry does through the research of chemists.[45]

The rhetoric in this quotation—that manufacturing chemistry was the application of pure chemistry to industrial processes—is similar to that used by English academic chemists and more or less echoes the last part of Kuhara's address. But Takamatsu did not necessarily mean this literally. Otherwise, his lecture would have defeated his purpose of instituting a separate program of manufacturing chemistry. Indeed, Takamatsu shortly afterward used this rhetoric to plead for the specific need to educate chemistry-related manufacturers and foremen at the tertiary level in the principles of chemistry "in order to avoid financial losses caused by industrialists' investment in chemistry-related manufactures without a thorough knowledge of chemistry."[46]

His true intension becomes clearer as he introduces his learning experience with Watson Smith as a model for this kind of chemical education. He presents the example of the Certificate in Technological Chemistry at Owens College Manchester instituted in 1881, which, according to Takamatsu, was designed along the lines of that of the Zurich Polytechnic in Switzerland. Takamatsu explained in some detail the three- or four-year curriculum of the Certificate in Technological Chemistry at Manchester consisting of pure and manufacturing chemistry, physics, mechanical engineering, geology, mineralogy, botany, and mechanical drawing.

In fact, as Robert Bud, Gerrylynn Roberts, and James Donnelly pointed out, engineering subjects were not in the scheme of the Certificate in Technological Chemistry at Manchester.[47] However, the curriculum for the Technical Branch of the School of Chemical Technology at the Zurich Polytechnic included the General Theory of Machinery, an equivalent of mechanical engineering.[48] One possible interpretation is that Takamatsu acquired information about this curriculum not from the published syllabus in college calendars, but from personal conversation with Watson Smith, who possibly referred to his own interpretation of the scheme, which reflected Smith's experience at Zurich more than the published version. A more important point is that the inclusion of engineering subjects would strengthen Takamatsu's argument for a separate teaching program for manufacturing chemistry.

Takamatsu then turned to the situation in Japan. He first mentioned that the Tokyo Shokkō Gakkō already had a secondary-level course for manufacturing chemistry (chemical technology) and that the Department of Chemistry at Tokyo University had already delivered lectures on this subject. Takamatsu then argued for establishing a new separate teaching program on manufacturing chemistry for final-year students who had a wish to become chemical technologists within the department. He concluded his address by calling for solidarity between manufacturing-oriented members and members specializing in "advanced" (*kōshō naru*) pure chemistry within the

Tokyo Chemical Society because "manufacturing chemistry is a branch of pure chemistry."

To understand Takamatsu's address, it is important to note that it was originally meant to be a congratulatory speech at the fifth annual meeting of the Tokyo Chemical Society, which had been established in 1878 with Kuhara as its first president. To celebrate this anniversary, the society invited executives of Tokyo University, such as President Katō, as well as high-ranking officials of the Ministry of Education. Among the latter was Hamao Arata, former vice director of the Tokyo Kaisei Gakkō under Hatakeyama and the current head of the Specialist Education Bureau (*Senmon Gakumu Kyoku*) who was in charge of Japan's higher education policy.[49] As Hamao was the main mastermind of the establishment of the Tokyo Shokkō Gakkō, Takamatsu had good reason to mention this school and differentiate his case from it in his address.[50]

The central message of Takamatsu's address, an argument for a separate teaching program of manufacturing chemistry within the Department of Chemistry of Tokyo University, was therefore too clear to be missed by guests from Tokyo University and the Ministry of Education. As the Ministry of Education's proposal to hire Wagener shows, the Ministry of Education in this period tended to emphasize the primary importance of on-site knowledge of the trade and practice of chemistry-related manufactures for training of chemistry-related manufacturers, so Takamatsu did not need to reiterate this. Thus, he purposefully took the opposite direction and emphasized the chemico-scientific (as opposed to engineering or on-site) side of the chemical industry by deploying the rhetoric that "manufacturing chemistry is a branch of pure chemistry" to make a case for a full-fledged course of manufacturing chemistry within the university.

Takamatsu also recognized another advantage of using the rhetoric of pure and applied chemistry—namely, that it would serve to unify scientific and manufacturing-oriented members of the society, a feature that was to be expected in congratulatory speeches to Japanese societies. Takamatsu's usage here of "*junsei kagaku*," which literally means "pure and right chemistry" carried the connotation of flattering pure chemists like Sakurai and Kuhara, the founder of the society and therefore the center of attention in this celebration, more than Kuhara's own "*junsui kagaku.*" Takamatsu's usage predates Sakurai's first public utterance of the same phrase in 1885.[51]

Takamatsu's use of the Western rhetoric of "manufacturing chemistry is a branch of pure chemistry" was closely related to another problem faced by those working for the institutionalization of technological education in chemistry in Japan during this period. Takamatsu had to overcome the class-based dislike or indifference toward working for private business on the part of students of samurai origin. For example, in 1883 a graduate of the Department of Chemistry at Tokyo University was offered, on Takamatsu's recommendation, a position as engineer at the Tokyo Metropolitan Gas Bureau (*Tokyo-fu Gasu Kyoku*) but resisted taking it when he heard of a plan for its privatization as the Tokyo Gas Company. Shibusawa Eiichi, the head of the bureau, and Takamatsu had to persuade the student to take the position.[52]

Historians unequivocally consider Shibusawa *the* key person in Meiji Japanese business history.[53] He had been born into a family of wealthy farmers and received a traditional preliminary education based on Chinese Confucian classics before entering the family businesses of farming, sericulture, and indigo manufacture. He later served the Tokugawa Shogunate as a financier and went to Europe to join the Shogunate mission to the Paris International Exhibition between 1867 and 1868. Shibusawa's experience of Western culture, especially how businesses worked and the higher status of businesspersons in Europe, was a revelation to him. After working at the Ministry of Finance of the new Meiji government for a short time, Shibusawa turned to business and was instrumental in introducing the Western joint stock company system into Meiji Japan, and he was also involved in the establishment of countless companies, including chemistry-related factories, such as the Tokyo Gas Company and the Tokyo Fertilizer Company (*Tokyo Jinzō Hiryō Kaisha*).[54]

To raise the prestige of businesspersons and entrepreneurs in Japan, Shibusawa felt it necessary to coin a Japanese word for entrepreneur, "*jitsugyō-ka*" as distinct from "merchant" in the traditional sense (*shōnin*), thereby translating Western "business sense" into the Japanese context of Confucianism. A key idea was presented in his discourse on "the identity of morality and economy (*dōtoku keizai gōitsu setsu*)," using rhetoric from Chinese Confucian texts to justify moneymaking businesses by regarding them not as selfish private affairs, but as national affairs so long as the business was reasonable and fair. In Shibusawa's own words, his close association with Takamatsu began with the above incident in 1883 and was due to the similarity between Shibusawa's discourse on the identity of morality and business and Takamatsu's own ideal of the complementarity of scholarly theory (*gakuri*) and enterprise (*jigyō*). Shibusawa thought that his ideal could be realized by Takamatsu's teaching activities.[55]

Takamatsu's close connection with Shibusawa therefore reveals a significant sociocultural dimension of his rhetoric of pure and applied chemistry discussed earlier. It had the function of justifying the existence of university-educated chemical technologists in the private sector, just as Shibusawa's emphasis on the complementarity of "theory and enterprise" and "morality and business" had for Japanese entrepreneurship. Their common family background of a wealthy farming family and a traditional education in Confucianism partly explain their similar outlook on moneymaking business (chapter 3).

Takamatsu was often described by his colleagues as "docile," "modest," "patient," "avoiding quarrels," and even "lacking courage."[56] On the surface this might have been how he appeared. However, he would defend his point of view with determination if necessary although this might involve skillful adjustment of what he had learned in England and Germany to deal with the current situation of his institution as well as Japanese cultural codes.

This approach contrasted with that of Sakurai and Kuhara, who pursued the strategy of "literal translation," i.e. rendering Williamson's or Remsen's ideal of pure chemistry into Japanese without modification. Takamatsu indeed succeeded in accomplishing the objectives expressed in this speech in a matter of months when he was promoted to the rank of professor.

The Separation of Pure and Applied Chemistry at Tokyo University

In August 1883, the fourth (i.e. the last) year of the curriculum of the Department of Chemistry was separated into pure and applied chemistry components (see table 4.2), and this consequently split the department into two: a Department of Pure Chemistry (*Junsei Kagakuka*) and a Department of Applied Chemistry (*Ōyō Kagakuka*).[57] Takamatsu was the instigator of this separation, as Takamatsu himself stated in his annual report to the president of the university: "[With] your sanction I divided the subjects of fourth-year students and made two departments, of pure and of applied chemistry."[58] No one else in the faculty made this statement in university documents available today.

Takamatsu made a nominal concession by agreeing to pick the term "applied chemistry" but in fact gained a great deal from this separation. The engineering side of chemical education for manufacturers was reinforced along the lines of the educational scheme for technological chemistry he had observed at Owens College. Mechanical drawing, mechanical engineering (applied mechanics), and a special laboratory course in manufacturing chemistry were introduced into the curriculum of the fourth year for students of applied chemistry.[59] The annual report of Sekiya Kiyokage, professor of mechanical engineering in charge of the lectures of applied mechanics for applied chemistry students, made this point:

> Students specializing in manufacturing chemistry have to make constant use of machines once they start to work on-site [in factories]. They obviously need knowledge of mechanics there. That is why students of manufacturing chemistry should take this subject.[60]

Takamatsu's own course on manufacturing chemistry (chemical technology) took the same approach, with "on-site" (*jitchi*) as a keyword. It examined the selection of raw materials and factory processes and studied theories of manufactures and the examination of products with the help of drawings of manufacturing machinery.

After the lectures, Takamatsu conducted tours for the students of firms such as sake and beer breweries and soap and gas factories in Tokyo to investigate the machines and processes on-site. A considerable proportion of every graduate thesis in applied chemistry supervised by Takamatsu was devoted to on-site investigation in workplaces, dealing with topics such as

Table 4.2 Curricula of the Departments of Pure and Applied Chemistry, Tokyo University, 1883–1884

	First year	Second	Third	Fourth
Mathematics (Analytical Geometry)	o			
Geometrical Drawing	o			
Mechanical Drawing				o (Applied)
Elementary Mechanics	o			
Applied Mechanics				o (Applied)
Astronomy (Outline)	o			
Physics	o	o	o	o (Pure)
Elementary Mineralogy	o			
Elementary Geology	o			
Mineralogy		o		
Metallurgy			o	
Logic	o			
English	o	o		
German		o	o	
Inorganic Chem. (Lectures + Lab)	o			
Organic Chem.	o			
Chem. Philosophy				o (Pure)
Study on Organic Compounds				o (Pure)
Chem. Technology			o	o (Pure/Applied)
Chem. Technology Lab			o	o (Applied)
Analytical Chem. (Qualitative Analysis)		o		
Analytical Chem. (Quantitative Analysis)			o	
Analytical Chem. (Assaying)				o (Pure/Applied)
Blowpipe Analysis			o	
Pure Chem. Lab				o (Pure)
Graduating Thesis				o (Pure/Applied)

Sources: Jpn. Calendar Tokyo Univ. Law, Sci., and Lit.

the "manufacture of Japanese paper," the "examination of rapeseed oil and its application," and the "dyeing of Japanese red-colored brocade."[61]

Takamatsu also started to develop a network of local manufacturers at that time. In July 1883 he was sent by Tokyo University to several provincial prefectures as well as to Osaka and Kyoto for "chemical experiments [*kagaku shiken*]."[62] On this occasion Takamatsu visited a local dyer, a weaver, a brewer, and a potter as well as sugar, vinegar, and soap manufacturers.[63] In the summer of the following year, Takamatsu went to the Kantō province "to collect samples for analysis," most probably for industrial purposes.[64] Takamatsu thus followed a decade later in the footsteps of his former teacher, Atkinson, by building a social network with indigenous manufacturers.

It is fair to say that Sakurai also benefited from the departmental split. Sakurai's course on chemical philosophy survived as one for fourth-year students of pure chemistry.[65] With the introduction of calculus and advanced physics into the curriculum for both pure and applied chemistry students, his desire to raise the level of physical and mathematical training of chemistry students was largely realized between 1883 and 1885 as he proudly reported in his 1885 presidential address to the Tokyo Chemical Society.[66]

Kuhara's increased involvement in departmental teaching was also closely connected to this split. Upon his promotion in 1883 Kuhara inaugurated the special lecture course on "research methods of organic bodies [*yūkitai kenkyūhō*]" for fourth-year students of pure chemistry. This course was for students who already knew the basics of organic chemistry, and it featured well-known reactions, theories on the structures of organic compounds, and preparation methods.[67] At the same time, Kuhara also started to supervise the graduate research of a pure chemistry student and assigned his earlier research topic on a red coloring matter extracted from a plant root with the object of investigating "its chemical properties and its structure, etc."[68]

It should be clear by now that, though Takamatsu was the instigator of these curricular changes, they were carried out in coordination with his colleagues such as Sakurai and Kuhara. Studies on Japanese chemistry in the 1880s by Tanaka Minoru, Hirota Kōzō, Furukawa Yasu, and myself have tended to emphasize conflicts and disputes between pure and applied chemists inside Tokyo University and the Tokyo Chemical Society, especially between Sakurai and Nagai.[69] However, the early use of *junsui* or *junsei kagaku* at the Tokyo Chemical Society and the split of Tokyo's Department of Chemistry, both on the initiative of Takamatsu, suggest what we might call the management of potential conflicts rather than a conflict itself.

Not all differences between Matsui, Sakurai, Takamatsu, and Kuhara's opinions concerning the educational philosophy of chemistry were solved by such "management": A major victim of conflict was Kuhara's plea against separating the teaching of pure chemistry from that of manufacturing chemistry. Differences between Takamatsu and Matsui's idea of applied chemistry would indeed erupt into a real conflict after the establishment of a Department of Applied Chemistry at the Imperial University in 1886 (chapter 7). However, Takamatsu's maneuver largely prevented such disagreements from developing into serious disputes. Through the separation of pure from applied chemistry in the Department of Chemistry, both parties largely gained what they wanted.

Toward the Establishment of the Imperial University

Important as it was, the separation of pure and applied chemistry at Tokyo University itself involved only the fourth and final year of the course in the Department of Chemistry. However, it was followed by a similar move on a much larger scale, that is, the separation of the Faculty of Industrial Arts (*Kōgeigakubu*) from the Faculty of Science in December 1885. The Department of Applied Chemistry, led by Takamatsu and Matsui, joined other engineering departments of the Faculty of Science of Tokyo University, such as those of mechanical engineering, civil engineering, and mining and metallurgy to form a separate Faculty of Industrial Arts. This soon merged with Imperial College of Engineering, Tokyo to establish a constituent College of Engineering (*Kōka Daigaku*) of the Imperial University in March 1886. The Department of Pure Chemistry, led by Sakurai, became part of the new College of Science (*Rika Daigaku*) of the Imperial University.[70] In short, this series of events led to the establishment of separate institutional entities for higher scientific and technological education in chemistry in Japan. Who was the originator of this seismic change?

Unlike the separation of the pure and applied chemistry departments, the establishment of a Faculty of Industrial Arts was a government initiative.[71] The Ministry of Education proposed the separation plan (without indicating how the various departments would be set up) to the Dajōkan on December 5, 1885. Ten days later, with Dajōkan's sanction, the Ministry simply notified Tokyo University that this would happen and indicated that the departments of mechanical engineering, civil engineering, mining and metallurgy, and applied chemistry would belong to the new faculty.

In its proposal to the Dajōkan, the Ministry of Education justified its plan in terms of the dichotomy between pure and applied science:

> In the Faculty of Science [of Tokyo University], not only pure science [*junsei no gakujutsu*] but also industrial applied science [*jitsugyō ōyō no gakugei*] is taught and investigated. Today, it is our urgent business to develop industries and lay the foundation of our wealth, and we must investigate industrial applied science and train industrial scholars [*jitsugyō gakushi*] for this purpose. It is urgently needed to institute another faculty to expand industrial applied science because the teaching of pure and applied science has different objects and methods. Therefore, I propose to separate industrial and applied-oriented departments from the Faculty and Science to institute a Faculty of Industrial Arts and to add other needed subjects so that we can expand

> industrial applied science [in this faculty] and greatly contribute to industrial development [in Japan].[72]

There are virtually no archival sources on the decision making within the Ministry of Education in the period leading to the establishment of the Imperial University in March 1886. Therefore, one has to infer the process from fragmented evidence and from knowledge of decision-making processes in the ministry.[73]

First, Hamao was the senior ministry official responsible for the policy of higher education as head of the First Bureau of Educational Affairs (*Gakumu Ichi-kyoku*, which was formed by reorganizing the Specialist Education Bureau) throughout this period. Nakano Minoru has clarified that in April 1885 Hamao discussed the reorganization of Tokyo University, including the planned merger with the Imperial College of Engineering, Tokyo with Ōki Takatō, who was then head of the Ministry of Education.[74] However, as Nakano pointed out, their discussions most probably did not go as far as suggesting a dividing up of the Faculty of Science at this stage.[75] Indeed, Hamao was absent from Japan on an official visit to Europe during the critical period between November 1885 and August 1886 and was not in a position to be directly involved with the reorganization. Who was, then?

The proposal quoted above was made in the name of Ōki, who was the head of the Ministry of Education (*monbukyō*) twice, first between 1871 and 1873 and again from 1883 to 1885. As discussed in chapter 2, during his first period in office as head of the Ministry of Education, Ōki was active in the reform of the Nankō and the government's overseas study scheme as well as in the establishment of the Education Law (*Gakusei*) of 1872. However, in his second period in office, he was overshadowed by Mori Arinori, one of the former Satsuma students at UCL, who was appointed commissioner (*goyōgakari*) of the ministry in July 1884 upon the strong recommendation from Itō Hirobumi, who had been a student at UCL at the same time as Mori. On December 22, 1885, Mori Arinori became the first Education Minister in the new cabinet system with Itō as prime minister. Also important to the above question is that Mori articulated the idea of separating the teaching of "pure" and "applied science," almost in identical terms as in the proposal, more or less during this time period.

Mori, "Applied Science," and Professional Training

Mori's idea was expressed in *Gakusei yōryō* (Essentials of government education policy), an undated memorandum that survived in the form of two drafts as well as a final version. Biographers of Mori have dated it to the end of 1885 or earlier in his term as commissioner.[76] In the final version of *Gakusei yōryō*, Mori argues that

> Science [*gakumon*] is divided into two divisions, pure science [*junsei-gaku*] and applied science [*ōyōgaku*], and both should be regarded as

> essential for the nation. The division of pure science can be small, but the division of applied science should be large enough.
>
> The object of [the division of] pure science is to pursue the truth of things. Therefore, the following two types of scholars should belong to this division and study [its specialties]: scholars who intend to benefit humanity and teachers at the higher level who should be called great doctors [*dai-hakushi*].
>
> [The division of] applied science is exclusively for the training of people who perform practical professional duties. Therefore, industrial people who put their learning to practical use to enrich the nation and benefit the people or those who will get qualification for government officials should belong to this division and study [its specialties].[77]

His own annotation of the final version of this document makes it clear that he had intended to render in Japanese the English terms "pure science" and "applied science." In the two earlier drafts, Mori also argued that "in the present situation of our country, applied science should be prioritized over pure science," though this clause disappeared in the final version.[78]

As Ivan Hall has pointed out, it is important to understand that Mori interpreted "pure science" and "applied science" in broad terms. He never opposed the idea that scientific training should be a part of technological education. He understood "pure science" as scholarly pursuits in general and "applied science" as what we might call professional training.[79] Thus, the similarity between Mori's *Gakusei yōryō* and the proposal of the Ministry of Education for the institution of a Faculty of Industrial Arts is too obvious to be dismissed. Above all, both emphasized and prioritized the training of practical industrial personnel. As Nakano suggested, it seems reasonable to infer that the separation of the Faculty of Industrial Arts from the Faculty of Science came from Mori and that the Faculty of Industrial Arts was established as a potential partner for a merger with the Imperial College of Engineering, Tokyo.[80]

Mori is a well-researched figure in modern Japanese history, but the origin of his ideas about pure and applied science is still obscure.[81] Of relevance to this question is Mori's participation in the short-lived, yet influential group of pro-Western intellectuals, the Meirokusha between 1873 and 1875. This group embodied the outward-looking and reformist zeitgeist of the *Bunmei kaika* period in the early 1870s.[82] A wide variety of political, economic, and religious issues were discussed in the Meirokusha, but as its original name, "the society of science, arts, and letters [*gaku, jutsu, bun, shachū*]" indicates, science and technology were well within the remit of this society. It was in the Meirokusha that Nishi Amane, a close friend of Mori, used the word *kagaku* as a translation of the word "science" for the first time in 1874.[83] He also used *gakujutsu* to explain the practical nature of Western science: "*Gaku* and *jutsu* are different in their natures, but in some cases of so-called *kagaku* (science) these two are mixed together and inseparable, for example, in chemistry."[84] This, however, does not explain the origin of Mori's rhetoric regarding the distinction between pure science and applied science. Nor is it clear to what extent Mori's views were influenced by Nishi.

More important from my viewpoint is Mori's association with his teacher and landlord during his London days, Charles Graham. The only written record about their relationship is on the reverse side of Graham's photograph that was in Mori's possession: "*Rigaku-shi* [Science Teacher or Master] Gurēmu: [I] stayed with him and often called him Father" (chapter 1). Mori later lived in London again as Japanese minister to Great Britain between 1879 and 1884. It is possible, if not currently verifiable, that he might have reunited with or heard about Graham during this period when the latter was professor of chemical technology at UCL (chapter 1). This overlooked relationship invites inferences about Graham's possible impact on Mori's views on science, technology, and education.

I characterized Graham's view of technological education as a "modified version of Williamson's liberal science model" in the sense that Graham believed that pure scientific training should come first before technological subjects at colleges or universities in apprenticeships at factories (chapter 1). Graham's scheme of a state-run school in his 1872 proposal to Itō was indeed close to the liberal science model. However, its function was predominantly to turn out senior technologists as well as teachers, government officials, and junior technologists. He never mentioned scholarly pursuits as among his model's goals (chapters 1 and 2).

The point that Mori was most likely to have taken from Graham was therefore his emphasis on "professional training" at colleges and universities. It is an important clue, but Graham did not go so far as to advocate the separation of an institution for the training of chemical technologists from a Faculty of Science, as Mori did later in Japan. Where did this idea come from?

Mori, Watanabe Hiromoto, and the Merger Issue

This conundrum will, I believe, be resolved if one stops to consider Mori's actions simply as the manifestation of his ideas and sees them instead as his efforts to solve a particular problem he was facing by using his own intellectual resources

As noted above, senior officials in the Ministry of Education had already started to discuss a plan for the merger of the Imperial College of Engineering, Tokyo with Tokyo University by April 1885, but their discussion most probably did not go so far as to suggest dividing the Faculty of Science at this stage. The proposed merger initially encountered opposition from the Ministry of Public Works. Watanabe Hiromoto, who was then deputy administrative head of the Ministry of Public Works (*Kōbu Shōyū*) wrote in May 1885 that "the transfer of the Imperial College of Engineering, Tokyo to the control of the Ministry of Education would induce students to indulge in scientific theory [*gakuri*]....According to past experience, in order to prevent students' minds from forgetting practice [*jissai*], it is better to keep this institution under control of a ministry that puts theory into practice."[85] Watanabe here directly mentioned a ministerial difference, but there is little doubt that he also intended to imply an institutional difference between the Faculty of Science of Tokyo University and the Imperial College of Engineering, Tokyo.

Contemporary as Watanabe's statement was, it would be problematic for historians to accept it uncritically.[86] In this case, however, perceptions were as important as realities and stood out in the bureaucratic negotiations of such merger plans. In this context, it would seem that Mori masterminded the separation of a Faculty of Industrial Arts from the Faculty of Science in December 1885 to counter such arguments as Watanabe's that the Faculty of Science prioritized theory over practice and used this separation as a key to move forward with the merger of the Imperial College of Engineering, Tokyo with Tokyo University to form the Imperial University in March 1886.

Mori's appointment of Watanabe, the early opponent of the transfer of Imperial College of Engineering, Tokyo to the Ministry of Education, as the first president (*sōchō*) of the Imperial University seems to have come from the same motivation—to contain potential conflicts arising from the merger. Watanabe indeed became a firm supporter of Mori's education policy after 1886. Mori's political genius was, it seems, that rather than creating a whole new system from scratch, he added a finishing touch to the unification plan for the Japanese higher education system that his predecessors and colleagues in the Ministry of Education had initiated.

Watanabe, Professional Training, and Chemistry

In its early years the Imperial University developed largely on the initiative of Minister Mori and President Watanabe with a particular focus on professional training at university. A federation of subject-based colleges (*bunka daigaku*), the Imperial University was characterized by oligarchic management and tight government control.[87] Decision making by colleges on academic matters, such as curriculum amendment and degree examination, was in the hands of the university president, college deans, and university councilors (*Hyōgikan*), all of whom were selected and appointed by the Ministry of Education. This structure remained in place until the reforms carried out between 1892 and 1893 under the leadership of Education Minister Inoue Kowashi when the professoriate of each college was given a certain degree of autonomy by means of the introduction of College Senates (*Bunka Daigaku Kyōju-kai*).[88] This does not necessarily mean that there was no leeway for professors to act on their own, but the relationship between professors and government and university officials were substantially altered after 1886. At the very least, this university hierarchy makes it crucial to examine how Watanabe's handling of institutional matters affected chemical education and vice versa.

No chemistry professors were appointed University Councilors throughout the first ten years of the Imperial University. Therefore, these professors could not officially participate in the university's management before the reform of the early 1890s. However, this fact resulted in neither a lack of influence in the decision making nor in the neglect of their subject, for chemistry was perceived by Watanabe to be in the national interest.

One example is Watanabe's address delivered at the Annual Meeting of the Tokyo Chemical Society in April 1886. It expressed his view of chemistry as a discipline that "contributed to the improvement of the material world of our country."[89] Watanabe encouraged members of the Tokyo Chemical Society to "inform people in our country more about the benefits of chemistry." He suggested two methods for accomplishing this task. One was to show adults how to improve their professional activities "in both traditional and new occupations" because "we can motivate adults only by immediate profit." The other was to "familiarize children with scientific theory by entertaining them." Watanabe's address thus foresaw the social functions of both applied and pure chemistry for industrialists and teachers at the Imperial University, which he would develop in his mission statement for the College of Science in 1887.

It is not known whether Watanabe's idea was his own or an outcome of his discussions with Sakurai and Takamatsu, for example, who had made similar points in past meetings of the Tokyo Chemical Society. As he was initially trained as a Dutch-style medical doctor and later worked as an official of the Ministry of Public Works, Watanabe probably had some ideas about the practical use of chemistry, especially for medical and industrial purposes.[90] A key point, however, is that there were some communication channels between Watanabe and chemists outside of the rigid university hierarchy and that Sakurai and Takamatsu shared the idea expressed in Watanabe's address, an idea that was consistent with his deep concern as president of the Imperial University about the kind of manpower his institution would be producing.

Watanabe backed up his rhetoric of the social benefit of chemistry with funding for state-of-the-art facilities at the Imperial University. He decided that the chemistry laboratories in the brand-new and completely Western-style buildings of the Colleges of Science and Engineering, which were completed in 1888, should have coal gas for fuel and electricity for lighting.[91] For this purpose, as early as October 1886, Watanabe ordered Sakurai, Divers, and Takamatsu to discuss the possible installation of facilities for manufacturing coal gas and a steam engine for generating electricity.[92] As Takamatsu recalled, the use of spirit lamps as laboratory fuel in the chemical laboratories at Tokyo University had caused difficulties in laboratory practice, especially in repairing glassware. Matsui also had difficulty providing laboratory training without gas. Watanabe's action, therefore, brought practical benefits to chemical laboratory teaching in the university and further integrated the material culture of scientific pedagogy into Japanese university architecture. It is another testimony to education officials' attitudes toward scientific pedagogy and architecture that did not exist in isolation but as parts of their overall attitude toward Western material culture.

Watanabe characteristically took a down-to-earth approach to university administration and emphasized professional training rather than scholarly pursuit.[93] The introduction of the Regulation of Student Loans for Undergraduates (*Bunka Daigaku Gakusei Taihi Kitei*) strongly reflected Watanabe's concerns about university finance and the educating of manpower.[94] As noted above, financial concerns had motivated the Meiji government to restructure Japanese higher education in the 1880s. For this reason, the newly established Imperial University was required to reduce expenditures and to partially fund itself through its own income. The obvious first step was to cut personnel: Kuhara, for example, resigned from Tokyo's Department of Chemistry and took up an appointment as chemistry teacher at the First Higher School in 1886, a preparatory school educating entrants to the Imperial University.

Financial measures other than personnel cuts were to reduce government-funded student grants and to raise the amount of tuition, both of which made the recruitment of students difficult. Furthermore, former samurai students who dominated the student population in the early years of the Imperial University were traditionally accustomed to free school education and a stipend provided by the Shogunate or domains and were in most cases too impoverished during the Meiji period to fund their studies without stipend.[95]

In this context, Watanabe made the decision in 1886 to restrict government-funded undergraduate as well as postgraduate student grants to departments that "focus on a pure and profound study of principles," such as those in the Colleges of Literature and Science. For application-oriented departments, he introduced externally funded student loans. He sent more than two hundred letters to government ministries and private companies to request funding for student loans for such departments and the hiring of graduates and postgraduates after the completion of their studies. Watanabe's effort paid off to some extent. For example, in the case of the Department of Applied Chemistry and the College of Engineering, it was government ministries, such as the Ministry of Agriculture and Commerce, private companies with political connections (*seishō*), such as Mitsui, Mitsubishi, and Fujita-gumi, and private companies that had originated in government industrial enterprises, such as the Tokyo Gas Company (*Tokyo Gasu Kaisha*) and Shinagawa Glassworks (*Shinagawa Galasu Seizōjo*) that funded student loans.[96]

The externally funded student loans also had an effect on the "pure" Colleges of Science and Literature. In 1886 the Ministry of Education promised to provide no more than 30 places for student loans to train teachers for secondary and normal schools.[97] The ministry did not originally indicate which colleges its money was for, but most of the places went to the Colleges of Science and Literature.[98] This funding scheme for training teachers developed into the Scholarship Regulation of the Ministry of Education (*Monbushō Kyūhi Gakusei Yōkō*) in 1890 for the purpose of training teachers of secondary schools, for which only students admitted to the Colleges of Science and Literature were eligible.[99] As most external funding bodies required graduates to work for them at least for a certain period, this system was instrumental in building networks between the Imperial University and officialdom and private companies through graduates and also in determining the social function of each college.

To sum up, in spite of its oligarchic management, the Imperial University never neglected the voices of its chemistry professors. Other informal mechanisms indeed existed and were used by the professoriate to make up for the lack of a formal mechanism for participating in decision making in the Imperial University at this time. For example, professors held unofficial meetings in their colleges and submitted proposals to the University Council prior to the official establishment of College Senates in the 1893 reforms.[100] In addition, meetings of professors and heads of department were held regularly at the College of Science. Sakurai, as head of department, joined meetings with Yatabe Ryōkichi, professor of botany at the College of Science and one of the university professors in the inner circle of Mori, to discuss important matters, such as degree regulations, student admissions, the distribution of student grants and student loans, alterations in the curriculum, doctoral degree examinations, overseas dispatch of college publications, and the distribution of college budgets among departments.[101]

There is indirect evidence to suggest that professors had some measure of control over their departmental finances as well. The Sakurai Jōji Papers include an official notification received by Sakurai in June 1911 telling him that he was discharged as "accounting officer (*kaikei kanri*) of goods of the Department of Chemistry of the College of Science, Tokyo Imperial University."[102] Unfortunately, it is not clear when he was appointed and what powers he had with this office, but the very existence of such a role suggests some budgetary responsibility. His often-cited autobiographical statement that he diverted funds for laboratory teaching to his research would seem to corroborate this.[103] It is likely that the same was the case for the Department of Applied Chemistry of the College of Engineering. Within these frameworks, Sakurai and Takamatsu created their own pedagogical spaces that more or less implemented their visions of chemical education.

Conclusions

The period between 1880 and 1886 was crucial for defining higher scientific and technological education in chemistry in Japan. Several teaching methods were tried, and a variety of ideas about chemistry and its teaching were discussed *publicly* during this period, providing fertile ground for subsequent growth of science education. The effect of such trials and discussions therefore went far beyond the narrower question of university restructuring.

Also important is that Matsui, Sakurai, Takamatsu, and Kuhara adopted totally different strategies to turn their visions of chemical education into institutional blueprints and to develop extramural networks with other sectors of Meiji society. Matsui took a largely incremental approach to improve his lot in departmental teaching during this period. Sakurai and Kuhara made explicit their Westernizing stance and did "literal" translations of Williamson's and Remsen's ideal of pure chemistry. In contrast, Takamatsu skillfully adjusted and more freely translated what he learned in England and Germany in the context of the current situation of his institution and Japanese cultural codes. All strategies paid off to a certain extent, but it was mainly Takamatsu's ability to iron out differences that enabled these scientists to work out a compromise.

However, it is misleading to emphasize Takamatsu's or any other professor's role too much. For the most part, the professors succeeded when their efforts were in conformity with the thinking of university and Ministry of Education officials such as Katō, Mori, and Watanabe. They were all concerned with university finance and favored the teaching of actual industrial practices as a framework of applied science teaching and the training of industrial and educational personnel. It would be also misleading to consider expert teachers as the shadow of university and government officials because the ideas of the latter underdetermine the reality of university teaching and politics. What this chapter showed is the mutual interaction between both parties through different channels and in different "translation" styles.

However, what was "defined" here is little more than institutional skeletons. Just as the proclamation of the Imperial University Ordinance was merely a starting point for the construction of the Imperial University as an educational institution, Sakurai, Takamatsu, and Matsui would have to construct their teaching programs, spaces, and contact zones for pure and applied chemistry and develop their social networks further according to their models, which is the main theme of the next three chapters.

Chapter 5

Constructing a Pedagogical Space for Pure Chemistry

The establishment of the Imperial University in Tokyo in 1886 changed the academic landscape of chemistry in Japan.[1] The Department of Pure Chemistry of the Faculty of Science of Tokyo University became the Department of Chemistry of the College of Science of the Imperial University under the leadership of Sakurai Jōji, who was appointed its head.[2] The Department of Applied Chemistry of the Faculty of Industrial Arts, under the leadership of Takamatsu and Matsui, merged with the Department of Practical Chemistry at the Imperial College of Engineering, Tokyo to form a Department of Applied Chemistry of the College of Engineering at the new university. Equally important was the move of Edward Divers (with his assistant Haga Tamemasa) from the Imperial College of Engineering, Tokyo to the College of Science, not to the College of Engineering.

Between 1886 and 1899 (the year of Divers's resignation), therefore, Tokyo's Department of Chemistry was in the hands of an English chemist and an English-trained Japanese chemist, both of whom were exponents of the liberal science model and its ideal of pure chemistry. The department played a key role as a major reference point for the subsequent development of higher chemical education in Japan. Chapters 5 and 6 address the question of what kind of pedagogical regime Sakurai, with the support of Divers, constructed for the Department of Chemistry as the top of the hierarchy of chemical education nationwide. I shall in particular look at the resources they could draw on and the constraints they had to satisfy. In doing so, I will take a somewhat unusual path of focusing on the pedagogical space first (chapter 5) and then examining the scientific pedagogy developed in this space (chapter 6).

The main reason why I take this path, of course, is my methodological concern with contact zones, spaces in which people with different backgrounds mingle, interact, and exchange things and ideas. The predominantly English professoriate interacting with Japanese students makes it natural to interpret the case in this way. Indeed, in addition to classrooms where Sakurai and Divers lectured in English or in a Creole mixture of English and Japanese (see chapter 6), the chemical laboratory was where students experienced what we might call culture shock. According to the recollection of Shibata Yūji (who entered the department as an undergraduate in 1904), Sakurai ordered students to behave appropriately when in the department building in an induction meeting. That is, students were asked to wear polished shoes and Western-style clothing in the chemical laboratories as much as possible.[3] Sakurai's instructions stemmed partly from practical concerns; he added that students who "refuse to wear Western clothes,...should use *tasuki* [a strip of cloth]" to prevent the long sleeves of their kimonos from hanging down dangerously, as some students actually did (see a student far behind in figure 5.1). Yet the vivid recollection of Shibata, together with his subsequent comment that "Professor Sakurai was an English gentleman and fastidious about clothes," strongly suggest that his students were quite likely to understand Sakurai's intentions culturally.[4] Classrooms and laboratories were really the place where students experience (for them) alien cultural practices, worthy of the name contact zones.

This episode, together with the accompanying picture, also nicely captures intricate power relationships between teachers, who tried to regulate and supervise, and students who occasionally defied laboratory regulations imposed by teachers in learning how to behave. I contend in this chapter

Figure 5.1. Student Laboratories, Department of Chemistry, College of Science, Imperial University, ca. 1894. Courtesy of the Department of Chemistry, School of Science, University of Tokyo.

that the supervision of students' work by teachers is key to understanding why Sakurai and Divers arranged classrooms, laboratories, office spaces, and other socializing spaces in a particular way and thus structured contact zones between teachers and students in a certain way.

Supervision, surveillance, and other social relations between those who control and who are controlled has long been the key interpretative focus of the Foucauldian analysis of architecture. It was inspired by the panopticon, the prison with a tall watch tower surrounded by cells, conceived by English utilitarian philosopher Jeremy Bentham.[5] More recently, Gregory Bracken started from Bentham and Foucault but departed from the rigid dichotomy of "controllers" and "controlled" in analyzing the Shanghai Alleyway House as "the benign panopticon." By this he meant that this housing typology enabled every resident to be a watcher and to form a vibrant interactive street life.[6]

Historians of science and science education adopted a similar way to interpret laboratory design. William Brock, for example, pointed out that the design of teaching laboratories at Finsbury Technical College in London, established in 1883, was affected by teachers' demand to make the supervision of students easier.[7] Using the same case, Graeme Gooday contextualized a contemporary debate on the supervision of students in laboratories in the process of legitimization of an academic laboratory space for electrical engineering in late nineteenth-century Britain.[8] Jeffrey Johnson argued that the design of chemical laboratories of the Universities of Bonn and Berlin in mid-nineteenth century Germany by August Wilhelm Hofmann reflected his concern about monitoring the laboratory work of students and assistants.[9]

Taking these insightful interpretations as a lead, my discussion will take us to a different conclusion: the dichotomy between controller and controlled is problematic. It is also different from Bracken's much more egalitarian "benign panopticon." Here, students were indeed controlled and so were teachers by the pedagogical space as contact zone once they started to work there. This insight justifies focusing on pedagogical spaces *as an agency* in the analysis of scientific pedagogy: without spatial analysis it is impossible to explain the development of teaching and research activities at Tokyo's Department of Chemistry.

"Red-Brick Fireproof Chemical Laboratory after a German Model"

Apart from foreign-language lectures and "outlandish" laboratory regulations, the building would also augment students' sense of culture shock. The two-storey Main Building of the College of Science (hereinafter abbreviated as "the Main Science Building"), which accommodated the laboratories and classrooms of the Department of Chemistry on the second floor and the Departments of Physics and Mathematics on the first floor, was a simple but elegant Renaissance-style brick building with slate-tiled roofs (figure 5.2).[10]

To understand the origin of this building, we have to return to the period just before the establishment of the Imperial University in 1886 discussed

Figure 5.2. Main Science Building, Tokyo Imperial University, ca. 1900
Source: Ogawa Kazumasa 1900. Courtesy of the National Diet Library, Japan

in chapter 4. The key characteristics underlying chemical or, more generally, scientific and technological education in early Meiji Japan in the 1870s was the heterogeneity of its higher education system (see chapter 2). Not only did several state schools for scientific and technological education stand side by side in this period, such as Tokyo University and the Imperial College of Engineering, Tokyo, there were also inconsistencies within a single institution. Tokyo University had two antecedent schools, Tokyo Medical School (*Tokyo I-Gakkō*), which formed its Faculty of Medicine, and Tokyo Kaisei Gakkō, which formed its Faculties of Law, Science, and Literature. The two retained different personnel, different teaching languages, and different campuses after they technically merged into Tokyo University in 1877 until they came under a single administrative body in 1881. Even then they continued to publish different calendars. The Faculty of Medicine was located on the Hongō campus, where Dutch and later German teachers dominated the teaching, and its lingua franca was German. In contrast, English had been the most popular language at the Tokyo Kaisei Gakkō, which occupied the Kanda-Nishikichō campus, since the early 1870s and became its only teaching language in 1874. American and British teachers therefore reigned in the Faculties of Law, Science, and Literature.

That was to change in the 1880s, however, when the government-controlled Japanese higher education system underwent a complicated restructuring process. An important event in this restructuring was the geographical transfer of the Faculty of Science of Tokyo University in the summer of 1885 from the Kanda-Nishikichō to Hongō campuses so as to share the site and buildings of the Faculty of Medicine (chapter 4). The building we are looking at in figure 5.2 was originally designed in 1885 as a chemical laboratory (*kagaku jikkenjō*) to be shared by all chemistry-related departments (pure, applied, and pharmaceutical chemistry) on the extended Hongō campus of Tokyo University, where the Faculties of Medicine and Science had been newly collocated.

According to official documents, building an interdepartmental chemical laboratory was motivated by the same factor as the campus relocation, i.e., bureaucratic and financial, not scholarly, considerations and approved by the government on the grounds of "school economy."[11] However, the impact of this building plan on chemical education in Japan went far beyond school administration. The building differed radically from the preexisting examples in Tokyo University. Whereas the three chemical laboratory buildings in the older Kanda-Nishikichō campus were timber-built and accommodated only one laboratory each, the new general chemical laboratory was conceived rather as an all-embracing complex accommodating a variety of laboratories, a lecture hall, lecture rooms, and operation rooms.[12] It gave chemistry professors at Tokyo University the material basis to think for the first time about an all-embracing space for chemical pedagogy.

The design was in the hands of a Paris-trained architect from the Ministry of Education, Yamaguchi Hanroku; additional advice was provided by the Berlin-trained pharmaceutical chemist Nagai (Wilhelm) Nagayoshi.[13] As Yamaguchi had had no experience in designing laboratory buildings, we can assume that Yamaguchi relied on Nagai in many ways for this project. Nagai was a professor of chemistry in both the Faculties of Medicine and Science when he was involved in the design, but he had particularly strong connections to medical professors through their common German background. The choice of Nagai as the academic advisor of laboratory design thus makes sense in light of the strong presence of medical faculty at that time in the administration of Tokyo University.[14]

According to the biography written by Nagai's former student, Kanao Seizō, he designed the laboratory on the basis of his study and research experiences in Hofmann's chemical laboratory in Berlin and was very proud of the "red-brick fireproof chemical laboratory after the German model."[15] Nagai's pride is understandable. The German chemical institute and its building saw unprecedented development in the latter half of the nineteenth century as organic chemistry flourished.[16] This tide reached chemistry in Japan in the mid 1880s through the restructuring of its higher education system and perhaps as a result of the upper hand the medical faculty had in the administration of Tokyo University in the mid-1880s.

A comparison of Yamaguchi's plans of the chemical laboratory at Tokyo (figures 5.3 and 5.4) and those of Hofmann's chemical laboratory in Berlin (figure 5.5) shows the extent to which Yamaguchi and Nagai's design followed the Berlin example.[17] Apart from the unmistakable similarity in shapes, several key elements of Nagai's academic advice are based on his experience in Berlin. For example, the plans included special rooms for a variety of operations, such as balance rooms, spectroscopic (photometric) and blowpipe analysis rooms on the first floor, and gas analysis and combustion analysis rooms on the

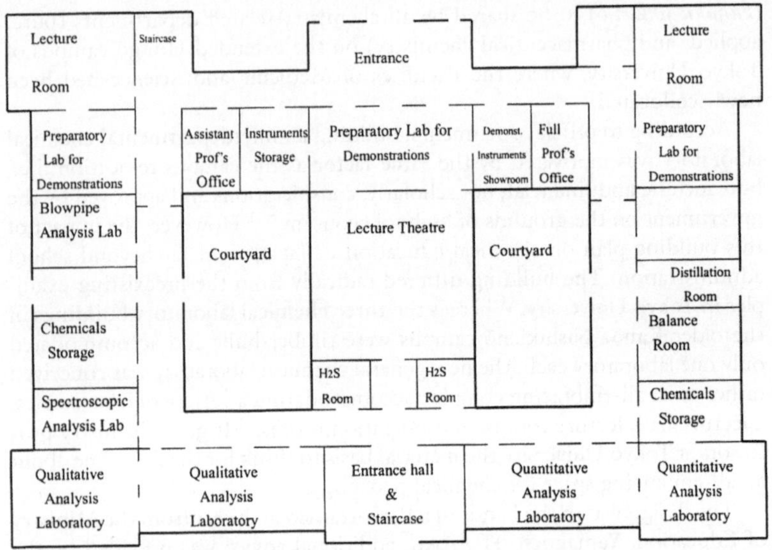

Figure 5.3. Drawing of the first floor of the "Chemical Laboratory," Tokyo University. Note that the direction toward the bottom is northbound. The author's drawing based on *Yamaguchi Hakushi Kenchiku Zushū*.

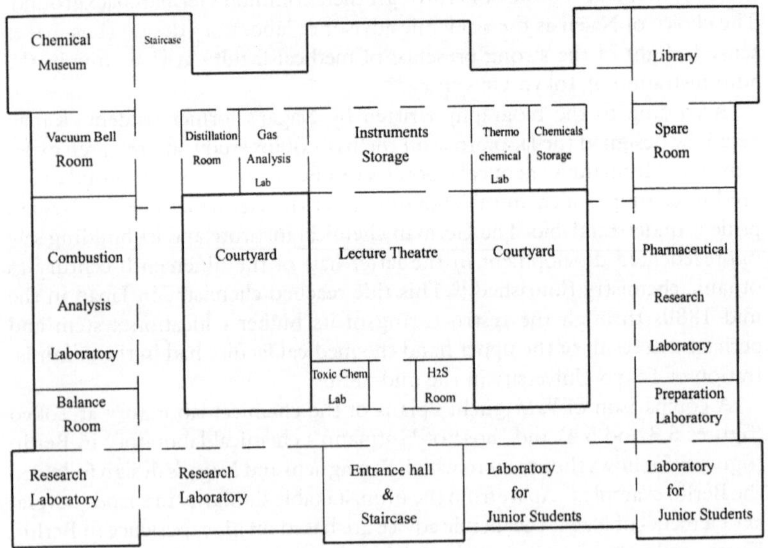

Figure 5.4. Drawing of the second floor of the "Chemical Laboratory," Tokyo University. The author's drawing based on *Yamaguchi Hakushi Kenchiku Zushū*.

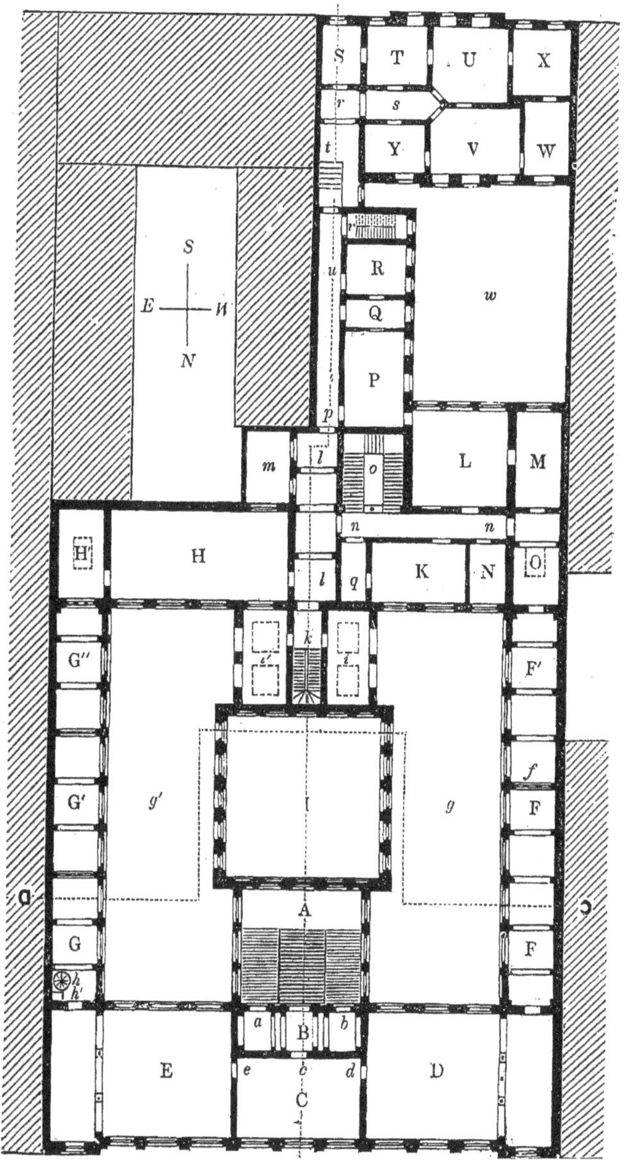

Figure 5.5. Architectural drawing of the second floor of the Chemical Institute at the University of Berlin. Note the Southern part of the building which was a residential space for the director and his family.

Source: Hofmann 1866.

second floor. There was also to be a chemical museum with showrooms for prepared chemicals and scientific instruments. The design also included preparatory laboratories for demonstrations next door to the central lecture hall and lecture rooms. All of these features had become commonplace in plans for chemical laboratories in Germany as well as in England by the 1870s, although in England they were not always implemented.[18] Second, Yamaguchi's plans contained only one office for a full professor and one for an assistant professor, which clearly reflected the one-chair-per-discipline system of German universities.

This last aspect of the laboratory design would seem not to have been suitable for Tokyo University, where several full professors of chemistry, including Sakurai and Takamatsu, had already been appointed at the time of Nagai's arrival in Tokyo in 1884. The multiple professorships in a single discipline at Tokyo date back to 1873, when William Eliot Griffis suggested to school administrators that they should hire three professors for the chemistry department at the Tokyo Kaisei Gakkō (chapter 2). As a newcomer to the university, Nagai failed to realize that the director-centered structure of his design would have clashed with the preexisting professorial structure of Tokyo University.

The most visible departure from the Berlin model was the absence of both a residence for the director and apartments for assistants from Yamaguchi's plans.[19] This difference involved the issue of urban design, of which Yamaguchi later became a leading expert in Japan.[20] The Japanese residences of most social classes were traditionally attached to workplaces up until the end of the Tokugawa period. Soon after the Meiji Restoration in 1868, however, political leaders in the new Meiji government started to use monumental government buildings exclusively in the Western architectural style and separated them from residences as a vehicle for the modernization-qua-Westernization of public administration to visibly distinguish their Meiji government from the Tokugawa Shogunate.[21] School architecture was right within this Westernizing trend. As a result, the independent residence, though still rare in Japan, functioned as a status symbol of high-ranking government officials in the late 1880s and began to spread among middle-class Tokyoites in the 1910s. In this sociocultural climate, there would have been no reason for Yamaguchi and Nagai to stick to the "older" style of attaching residential spaces to workplaces even if they were included in the German original.

Overall, Nagai and Yamaguchi did a good job of balancing their laboratory design between several subdisciplines of chemistry. For example, combustion analysis, the standard technique of elemental analysis throughout and even after the nineteenth century, was more likely to be used by organic chemists thanks to the improvement (and subsequent popularization) of this technique in 1830 by Justus Liebig.[22] Spectroscopic analysis was widely used by inorganic chemists to find "fingerprints" of chemical elements following the collaboration between Robert Wilhelm Bunsen, a chemist, and Gustav Robert Kirchhoff, a physicist, in 1859.[23] These techniques were quite versatile and a staple *any* chemist in the late nineteenth century should have learned at an early stage of his or her career. Nagai and Yamaguchi justifiably included spaces for both these techniques in their design of a "general chemical laboratory

It is therefore all the more striking that the Yamaguchi-Nagai plan deviated from the Berlin one in that the former included a "pharmaceutical research laboratory" on the second floor, whereas the latter did not have any laboratory for a particular branch of applied chemistry, whether pharmaceutical, medical, or industrial chemistry. Despite his close relationship with the dye industry through his former students, Hofmann at the University of Berlin identified himself strongly as an academic chemist and saw "no reason to treat my academic subject other than in a purely academic manner."[24]

Considering Nagai's specialty of pharmaceutical chemistry, it is reasonable to interpret the inclusion of the "pharmaceutical research laboratory" in Nagai's plan as a substitute for a director's private laboratory, which was absent from Nagai's plan due to the aforementioned omission of residential spaces. In other words, Nagai considered himself presumptive director of Tokyo's general chemical laboratory. Indeed, we can see from the position of the pharmaceutical research laboratory that, while doing his own research, Nagai would also have been in a good position to monitor the laboratory work of students and assistants. This had been the main concern of Hofmann in planning his laboratories, as mentioned at the beginning of this chapter.[25]

Despite a modification necessary in the early-Meiji sociocultural context, Nagai's laboratory plan closely followed the design of Hofmann's laboratory and showed a strong orientation toward a centralized hierarchy with control in the hands of the director. This was the main characteristic of German chemical institutes and reflected the one-chair-per-discipline system of the German university system. Construction of the laboratory began in November 1885, and the building was completed in December 1888.[26]

Redesign of the Chemical Laboratory by Sakurai and Divers

The reorganization of Tokyo University into the Imperial University in March 1886, driven by the same force of restructuring that had given rise to the building plan, completely changed the plan in two ways. First, the constituent subject-based colleges of the Imperial University were defined as institutions independent from each other (chapter 4). This put an end to collaboration between the Faculties of Science and Medicine and made intercollegiate facilities, such as a general chemical laboratory, unacceptable. Furthermore, allocating a costly building to only one discipline became also unacceptable, as professors of other departments were also lobbying for new buildings. For these reasons, at least officially, the Imperial University Council, the governing body of the university, decided in May 1888 to include the Departments of Mathematics, Physics, and the college administrative office in the newly built "chemical laboratory." This meant that the Department of Chemistry could occupy only the second floor of the building.[27] Second, Nagai, the originator of the idea of building a general chemical laboratory, lost his position at Tokyo University when it was reorganized into the Imperial University. As Nakano Minoru pointed out, the emerging Imperial University was under immense pressure from the government to reduce costs by cutting staff (chapter 4). This resulted in Nagai's dismissal and that of over 50 other members of the teaching staff, all of whom had been "off duty" (*hishoku*) at the time of reorganization in March 1886.[28]

When Divers, who moved to the Department of Chemistry from the Imperial College of Engineering, Tokyo as Sakurai's colleague, reported to the university for the academic year of 1886–1887 that "the new laboratories that Professor Sakurai and I have planned are now under construction," he meant that they had taken over the responsibility of designing the interior of the building from Nagai. This interior design would have included the furnishing of laboratories and lecture rooms in addition to the reallocation of rooms.[29] Indeed, Sakurai and Divers furnished them with equipment imported from England and had coal gas pipes installed to laboratories.[30] They could use their experiences in and connections to England in designing and appointing their new laboratories in Tokyo.

The collocation of the Departments of Chemistry, Physics, and Mathematics in one building meant two things to Sakurai and Divers. First, this opened up the possibility of chemists collaborating with physicists and mathematicians, which was definitely in line with Sakurai's teaching philosophy articulated earlier in the 1880s (chapter 4). Second, the number of laboratory rooms was reduced. Sakurai and Divers had to give up most of the special operation rooms, preparatory laboratories, and chemical museums planned by Nagai.

Nevertheless, a comparison between Yamaguchi's plan of the second floor of the building (figure 5.4) and the drawing of the space for Department of Chemistry (figure 5.6) shows that Sakurai and Divers assigned exactly the same spaces for student laboratories as Yamaguchi had designated, suggesting that Sakurai and Divers adopted the design of Yamaguchi and Nagai in this particular respect.[31] The photograph taken inside the laboratory (figure 5.1) shows that Sakurai and Divers adopted an alignment of working benches running parallel to the main, longer walls, presumably to make the most of natural lighting and ventilation through the large windows. This alignment had been used in Liebig's famous analytical laboratory at Giessen (built in 1840).[32] This design was later adopted in UCL's Birkbeck laboratory (built in 1846 loosely based on the Giessen model) where Sakurai was trained, in Hofmann's chemical laboratory in Berlin, and probably in the laboratory of the RCC, where Divers was trained and Hofmann was the first professor of chemistry between 1845 and 1865.[33] Such international circulation of laboratory design undoubtedly made it easy for London-trained chemists to adapt themselves to the plan based on the Berlin example.

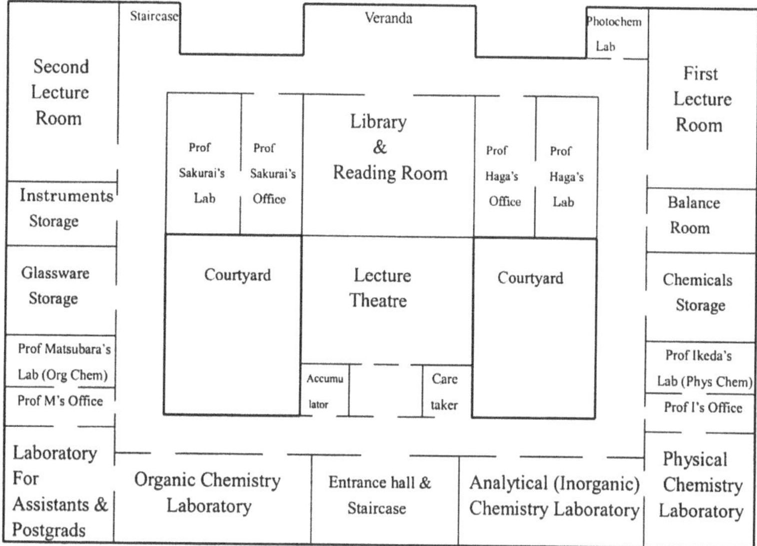

Figure 5.6. Drawing of the second floor of the Main Science Building, occupied by the Department of Chemistry. My drawing with translation based on Iimori Satoyasu's sketch in possession of the Department of Chemistry, School of Science, University of Tokyo.

However, the initial 1885 plan underwent important modifications related to the hierarchy, subject matters, and supervisory practice in the teaching of the department. First, the part of the Yamaguchi-Nagai plan that accommodated the one-chair-per-discipline rule was discarded, and both Sakurai and Divers were assigned one lecture room, one office, and one private laboratory of exactly the same size and form as those of a full professor. Second, instead of the binary division of students' laboratories into junior and advanced ones seen in Hofmann and Nagai's examples, Sakurai's laboratories were divided by subdisciplines of chemistry, i.e., organic, analytical (inorganic), and physico-chemical laboratories, corresponding to Sakurai's own teaching program (chapter 6). The students' laboratories filled the northern side of the quadrangular building in the drawing, together with a research room for assistants and postgraduates.

Division of Labor in the Pedagogical Space

The position of the professors' offices and private laboratories in relation to that of the students' laboratories are important features of Sakurai's and Divers' pedagogical space. Here, the plan of Sakurai and Divers deviated most from the original plan created by Yamaguchi and Nagai. According to the drawing, the two offices for senior professors, facing courtyards, were close to the lecture rooms but were physically isolated from the student laboratories. The only doors of the private laboratories of Sakurai and Divers led not to the hallway but to their contiguous offices and were functionally closed off from outside. This would have been an extremely awkward arrangement if Sakurai and Divers had had to monitor students' experimental work. No wonder that the student in a kimono in one of the student laboratories did not worry about the possibility of being spotted by Sakurai, who admonished students not to wear the traditional robe (figure 5.1).

In contrast, the two offices for junior professors were directly connected to the student laboratories. Students could enter the junior professors' offices and vice versa without using hallways. This arrangement virtually defined the respective roles of the senior and junior professors expected by Sakurai and Divers, who designed the entire floor. The supervision of students' experiments *at all levels* was basically the role of junior professors, whereas the main role of the senior professors in relation to the undergraduates was to prepare and deliver lectures.[34] Indeed, according to Sakurai, "though all experimental teaching is under the overall direction of Professor Divers and myself, day-to-day training was carried out by two assistant professors."[35] Divers also reported that Haga Tamemasa, one of two assistant professors, was in charge of supervising students' experiments.[36] This separation of the roles of senior and junior professors in the Department of Chemistry had the effect of making consecutive junior professors, Haga in the 1880s and early 1890s and Ikeda Kikunae and Majima Toshiyuki (or Rikō) in the late 1890s and 1900s, the central figures in laboratory teaching regardless of students' levels, while Sakurai and Divers concentrated on their role of senior professors as lecturers defined by the structure of this pedagogical space.

Sakurai almost certainly intended this separation on the basis of his experience of working with Williamson at UCL as discussed in chapter 3. I do not argue that the design of the Birkbeck Laboratory influenced this aspect of his design. With the common origin in Giessen, it had much in common with the Berlin laboratory, if not so grand and all-embracing as the latter. What mattered was Williamson's use of it. Sakurai took Williamson as a role model, emulating the "division of labor" structure of chemical education at UCL that I described in chapter 1 and constructing his pedagogical space accordingly with Divers, who had experienced the same division of labor while he worked as chemistry assistant at Queen's College in Galway.[37] In a letter to Williamson's daughter, expressing his indebtedness to her father, Sakurai commented that "I do not remember seeing [Williamson] engaged in any experimental work while I was at University College (1876–81)";[38] Williamson evidently rarely showed up in the laboratory despite having his own private laboratory and office located next to the student laboratory (chapter 3).[39]

In contrast, in the same letter Sakurai enthusiastically praised Williamson's greatness as a theoretical chemist and his systematic approach to lectures, proclaiming that "I think I got my habit of endeavouring at clearness and precision from attending lectures [of Williamson]"[40] If what Sakurai learned from his mentor mostly came from classroom lectures, it is hardly surprising that Sakurai saw classrooms as his principal contact zone with students and wanted to concentrate on preparing for lectures in a secluded office and private laboratory without being disturbed.

Students' recollections also corroborate this picture. They unanimously praised the clarity and order of Sakurai's lectures (chapter 6), but also stated that Sakurai rarely visited students' laboratories. Majima, for example, studied at Tokyo's Department of Chemistry between 1896 and 1899 and later worked there as an assistant professor until 1907. While praising Sakurai's lectures, he confessed his dissatisfaction with Sakurai's supervision of students' experimental work. As a graduate student under Sakurai, Majima received no advice about the selection or the execution of his research topic. He recalled that while he welcomed the freedom that Sakurai gave him, it also made him feel somewhat insecure.[41]

In contrast, Majima acknowledged in later years his indebtedness to Haga in his laboratory work. Trained by Divers thoroughly in chemical analysis, the staple of chemical pedagogy, Haga proved to be a versatile experimental trainer between the 1880s and early 1900s. For example, Haga coached both Majima and Ōsaka Yūkichi, who made their names as adept experimentalists in organic and physical chemistry, respectively. This is something of a surprise considering the specialization developed in chemistry during the nineteenth century. In later years both acknowledged their indebtedness to Haga who improved their experimental skills.[42] This division of labor also made for closer relationships between junior professors and chemistry undergraduates in laboratories, just as the supervisory practices at UCL and the Owens College Manchester had done for the relationships between Charles Graham and the Chōshū and Satsuma students in the 1860s and between Takamatsu and Watson Smith in the late 1870s.

The junior professors' role as experimental trainers was partly that of "cultural mediator," taking daily care of Japanese students who might have felt culturally uneasy because of the department's strongly Western atmosphere due to the predominantly English professoriate. Haga particularly suited this role. Apart from his friendly character, especially to youngsters, he had had a traditional educational background in Chinese Confucian studies as the son of a samurai before studying chemistry with Divers at the Imperial College of Engineering, Tokyo. Unlike Sakurai, Haga did not study overseas during the period of his character formation.[43] It is scarcely surprising then that Japanese students found him more accessible than either Divers or Sakurai.

Ikeda, Haga's successor as assistant professor in 1896, played a role as a trainer in glassblowing technique, another staple of chemical pedagogy, and physico-chemical measurements along the lines of Sakurai's physicalist approach to chemistry and chemical education.[44] According to Shibata's recollection:

> [Ikeda] sensei also had a distinct style as a teacher, and he was not the kind of professor who routinely delivered lectures and trained students

> in laboratories. Ikeda sensei drilled the essence of chemistry into the heads of freshmen in the form of volumetric analysis. Analytical technique in those days was not as manifold as it is today, and it was only in volumetric analysis that we used apparatuses at all. With simple graduated apparatuses like burettes and pipettes as a starting point, he freely explained the theory and practice of measurement, the concept and theory of errors, the method of calculation, and its results. He trained freshmen thoroughly until they understood his explanation by means of their own experiments.[45]

This quote captured a moment when Ikeda was introducing volumetric analytical chemistry revolutionized by physical chemistry and the use of mathematical error theory in treating data from quantitative analysis, just as contemporary physical chemists in American colleges and universities were doing.[46] Indeed, Ostwald's textbook on analytical chemistry, the second edition of *Die wissenschaftlichen Grundlagen der analytischen Chemie*, was translated into Japanese by Kametaka Tokuhei, a student of Sakurai and Ikeda, in 1899 with corrections by Sakurai.[47]

It was therefore not a coincidence that Tokyo's teaching laboratory for analytical chemistry was located next to its physico-chemical laboratory, which facilitated Ikeda's supervision. By the same token, being accommodated in the same building as the Departments of Physics and Mathematics meant that it was easy for chemistry students to take courses in physics (both lecture and laboratory) and mathematics and vice versa, and to network with students in these two departments, just as Sakurai was doing regarding his own physico-chemical researches with his physics colleagues, such as Cargill G. Knott and Yamagawa Kenjirō (chapter 6). Sakurai's allocation of laboratory rooms was thus carefully coordinated with how he intended to teach experiments in chemistry there.

Research Practice, the *Zasshi-kai*, and the Central Departmental Library

What stands out in my analysis of Sakurai's pedagogical space is the strong influence of the British "assistant-centered structure" at UCL and Owens College Manchester, where the junior faculty had an important role in training experimental chemists (chapters 1 and 3). One implication of this characteristic of Sakurai's pedagogical space was its effect on how research practice developed in Tokyo's Department of Chemistry. Its curriculum did not originally include a graduate thesis for the final-year students in 1886. However, spatially induced close relationships between junior professors and chemistry undergraduates in laboratories did occasionally lead to research partnerships, albeit as a by-product of undergraduate teaching rather than as the department's mainstream product.

For example, Haga's supervision of the experimental training of Majima and other students resulted in joint papers.[48] Ikeda also coauthored research papers with his students.[49] The most telling fact is that the annual Departmental Conference for the Presentation of Graduate Theses (*Sotsugyōsei Gyōseki Hōkokukai*), where final-year students had the obligation to present their graduate theses, was instituted in 1906 not by any of the senior professors, but on the suggestion of Majima, who was then assistant professor and had already started to look to Germany rather than to England as a model to follow.[50] The "research imperative" there moved upward from junior professors rather than downward from senior professors.

However, it does not follow that its senior professors, such as Sakurai and Divers, did not do anything to encourage research. The opposite is the case. The senior professors knew that doing original research is not just using manual dexterity in the lab, though that is a fundamental part of their métiers. The basic problem Sakurai and Divers faced was that chemical research for its own sake, without consideration of potential industrial applications, had scarcely existed before the establishment of the Imperial University in 1886. Sakurai and Divers helped students to *voluntarily* cultivate their "enquiring mind,"[51] something akin to what the English liberal education in the nineteenth century aimed at, and to engage with the frontiers of international research. They did so in classroom lectures and tutorials, including research supervision, in their offices (chapter 6). In addition, Sakurai also provided such research support in a more relaxed and informal atmosphere at the *Zasshi-kai* (the Journal Meetings) in the departmental library and its reading room, the third site of interaction between students and teachers that occupied the center of the pedagogical space.

These meetings were a departmental reading seminar, held once a week at the departmental library, where students and teachers alike made presentations based on papers in various Western chemical journals and then critiqued each other. Official departmental histories provide evidence of another aspect of the *Zasshi-kai*, namely, that of organizing social events for students and teachers to help them relax and to foster comradeship and solidarity in the department as a whole.[52] Sakurai was not the first Japanese chemist to introduce this type of gathering to his peers. His former classmate and colleague at Tokyo University, Kuhara Mitsuru, first publicly used the term *Zasshi-kai* (in Japanese) to explain his experience of attending Journal Meetings organized by his mentor, Ira Remsen, at the JHU (chapter 4). Sakurai perhaps heard about the idea of establishing a *Zasshi-kai* from Kuhara, but he also had a similar experience at UCL. He was a founding member and a contributor to its Chemical and Physical Society, which shared the same objectives and format as the *Zasshi-kai* (see chapter 3).

Why did the *Zasshi-kai* matter? The combination of solidarity and lively discussion, including criticism, to generate aspirations for original research, has been observed widely by historians in Japanese academic laboratories in the Meiji and Taishō periods. James Bartholomew points to this activity as a key to the management of the inherent tensions between the conformity of group solidarity in Japan and the creation of an innovative research atmosphere.[53] The *Zasshi-kai* at Tokyo's Department of Chemistry is an early example of this phenomenon that has hitherto been overlooked.[54]

The *Zasshi-kai* in fact originated in two separate organizations. One was a reading seminar, probably organized by Sakurai himself based on the JHU and UCL models. He mentioned this in "The Journal Meeting Minutes" in one of his notebooks.[55] The other was what the proceedings of the *Zasshi-kai* called the "Chemistry Alumni Society" (*Kagaku Dōsō-kai*), organized by students for the purpose of socializing and chatting.[56] Both were created in 1890 but struggled to survive in their first few years.[57] They began to function well only after Sakurai himself joined the Chemistry Alumni Society and invited students to his house or to other venues several times in 1893 for social gatherings. There Sakurai and his students enjoyed eating, drinking, informal chatting, singing, and playing *fukubiki* (the Japanese equivalent of raffle draws) and *shōgi* (the Japanese counterpart of chess).[58] The name "The Chemistry Alumni Society" was used in the proceedings of the *Zasshi-kai* until early 1897, but the society had merged functionally with Sakurai's Journal Meeting by March 1893.

From then on, the *Zasshi-kai* was organized by students in close cooperation with teachers such as Sakurai and Haga, but not Divers. The students held a variety of social events, such as parties and excursions, together with regular reading seminars. The aforementioned Departmental Conference for the Presentation of Graduate Theses was in fact another important scholarly event in which the *Zasshi-kai* was involved.[59] It functioned as one of the core social gatherings of the Department of Chemistry that fostered comradeship among students, graduates, and teachers.

This account makes it necessary to reassess Bartholomew's portrayal of Sakurai as an "aloof" professor.[60] In formal laboratory situations, Sakurai was not involved in day-to-day interactions with students. In his role as senior professor, he focused on lecturing rather than on direct experimental training, and his research was done on an individual basis. However, outside the formal academic structures, Sakurai had a pivotal role in making the *Zasshi-kai* a lasting departmental organization by giving it both a scholarly and social character. He accomplished this by associating with students outside the lecture room in an informal atmosphere. In other words, Sakurai succeeded in introducing his idea of using a reading seminar to raise Japanese students' aspirations toward original research for its own sake by appealing to the group solidarity so prevalent in Japan and to the students' leisure culture.[61]

The centrality of the *Zasshi-kai* in the pedagogical regime of Sakurai's Department of Chemistry is underlined by the central location of the departmental library and reading room in the departmental space, adjacent to the offices of two senior professors; we do not see this in either Hofmann's, Nagai's, or Williamson's laboratories. Perhaps the scarcity of Western books and periodicals made a specialized library all the more important for science professors and students in Meiji Japan. But the central location of the library is also due to the important role Sakurai ascribed to the *Zasshi-kai* in his transfer of "pure" chemical research as mental training to Japan.

Conclusion

In constructing a pedagogical space for Tokyo's Department of Chemistry, Sakurai had a couple of resources to draw on and constraints to satisfy: the American heritage of multiple professorships, his own experience at UCL, a new trend in laboratory design coming from German chemical institutes, and Japanese cultural elements brought by his students. Combining these elements was not an easy task at all and required much more than merely duplicating a blueprint or architectural style. Rather, Sakurai's task called for a creative process of distilling meanings from his learning experiences, from the laboratory designs he was familiar with, giving all of them the stamp of his own ideas, and from his interactions with students. Sakurai's partition of students' laboratories based on subdisciplines was a major departure from the mainstream laboratories in Europe and the United States and shows his forward-looking attitude toward chemical pedagogy emphasizing physical chemistry, a new discipline at the crossroads of physics and chemistry. Also, the precarious start and later success of the *Zasshi-kai* shows the importance of collaboration of teachers and students and Sakurai's skill in creating a workable contact zone.

Once completed, however, the design of the space of Tokyo's Department of Chemistry exerted strong constraining effects on pedagogy. It was essentially designed for a two-chair-system and simply could not catch up with the increase of departmental chairs that went hand in hand with the increase in the number of students (table 5.1). Another constraining effect was that it was impossible for lecturers to do demonstration experiments for lack of preparatory laboratories. Most important of all, Sakurai's design of pedagogical spaces virtually assigned completely different roles to senior and junior professors (regardless of their official titles) as lecturers and as experimental trainers, respectively. Indeed, this setup proved harmful to the career of Haga, the pivotal figure in the laboratory training of the Department of Chemistry in the 1880s and 1890s. He succeeded to Divers's chair in 1899 and had to completely change his role to that of lecturer, a circumstance he found extremely difficult to adapt to quickly due to his habit of stuttering.[62] Sakurai and Divers's pedagogical space failed to give Haga an opportunity to gradually remedy this drawback. It would not be possible to understand Haga's problem fully without spatial analysis of the scientific pedagogy at the department.

The above-mentioned constraining effects suggest a rather limited life cycle of Sakurai's design, and this was actually the case. The construction of a separate building for the Department of Chemistry near the Main Science Building, called the "Chemical Institute" (*Kagaku Kyōshitsu*), began in 1913 in the early Taishō period and was completed in 1916 (figure 5.7). The

Table 5.1 Student enrollment in the Department of Chemistry, College of Science, Imperial University, 1886–1900

	Post-graduates	3rd year	2nd year	1st year	Occasional	Total	Under-graduate Total
1886	0	0	0	2	1	3	3
1887	0	0	2	3	1	6	6
1888	0	2	3	1	0	6	6
1889	2	3	0	1	2	8	6
1890	2	0	1	1	3	7	5
1891	2	1	1	2	4	10	8
1892	1	1	2	2	4	10	9
1893	2	2	1	5	4	14	12
1894	3	1	3	11	1	19	16
1895	2	3	8	8	1	22	20
1896	3	7	4	8	1	23	20
1897	5	4	6	5	0	20	15
1898	5	6	4	3	1	19	14
1899	9	3	3	2	1	18	9
1900	8	3	2	4	1	18	10
Average						13.53	11

Source: Jpn. Calendars (Tokyo) Imp. Univ.

entire department was relocated to this building in the same year.[63] The new chemistry building was L-shaped, making future extensions easier; it was built with ferroconcrete and red brick and stone facing and designed by Yamaguchi Kōkichi, another architect from the Ministry of Education, with academic advice from Ikeda.[64]

In Ikeda's hands, the departmental space underwent two major changes. First, the sharp distinction between classroom and laboratory teaching was blurred by the reintroduction of a preparation laboratory for lecture demonstrations next to the lecture room located at the western end of the L-shape on the second floor (at the northern end of which a student laboratory was located). Second, the hierarchical structure of Sakurai's pedagogical space, which underpinned the division of roles between senior and junior professors, was all but abolished. In the new building with larger available areas, all offices and private laboratories of full and assistant professors were equally located around the center of each floor whereas lecture rooms and student laboratories were at the both ends of the building.[65] In short, the location of professors' offices and laboratories no longer defined their roles in pedagogy.

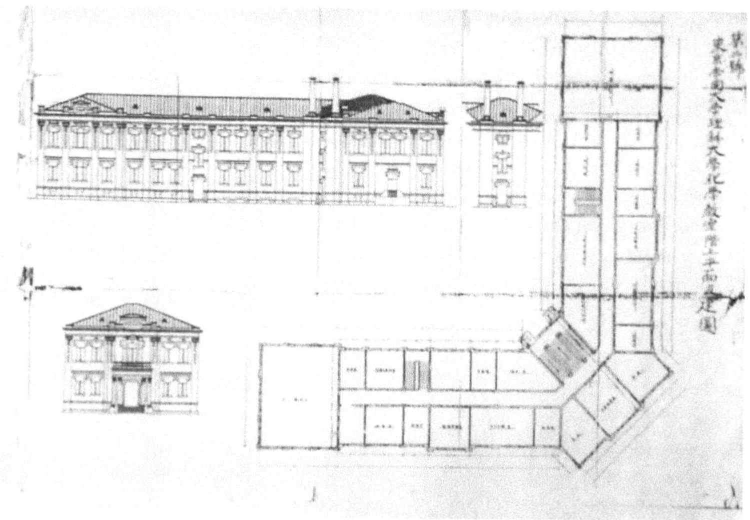

Figure 5.7. Architectural drawing of the second floor of the new Chemical Institute, College of Science, Tokyo Imperial University. Courtesy of the Department of Chemistry, School of Science, University of Tokyo.

Existing sources do not tell why Ikeda designed the new Chemical Institute at Tokyo this way, but one could infer his motive by means of an informed interpretation of his design. The building he would have most closely examined during his overseas study in Germany with Ostwald in 1899–1901 was the Institute for Physical Chemistry at the University of Leipzig, which was completed in 1898. It had a lecture hall with a preparatory laboratory at one end of the building, and this likely inspired Ikeda.[66] But the Leipzig example in its original structure did not have an L-shape, and as a German institute, it still had a director-centered structure, something that was alien to Ikeda's own design.

Ikeda's prior experience at Tokyo mattered as well. In the older building, Ikeda had been forced to continue occupying the place of a junior professor as an experimental trainer in glassblowing and analytical and physical chemistry long after his promotion to full professor in 1902. These circumstances lead me to regard Ikeda's design as a sign of his self-assertion as a chemical pedagogue and of his frustration experienced in Sakurai's hierarchical pedagogical space. While faithfully following Sakurai's "physicalist" approach to chemistry and chemical pedagogy as his former trusted student and brother-in-law, Ikeda largely behaved as a dissident in laboratory design. This new building mainly served Ikeda himself but not Sakurai (retired in 1919), and it served Shibata who succeeded Haga (resigned in the same year due to illness) in 1913 to the chair of inorganic chemistry, and the succeeding generations of professors in the Department of Chemistry.

However, one element of Sakurai's pedagogical space, the centrality of the departmental library, outlasted the buildings life cycle. Reflecting the active and privileged existence of the *Zasshi-kai*, the departmental library, which consisted of one reading room and two book storages, was situated at the center of the L-shaped Chemistry Building on the second floor above the magnificent façade, which only illustrious guests of the department could pass.[67] Thus, laboratories and their spaces show a generational divide, coincidentally between the transition period from Meiji to Taishō, as well as continuity in the development of Tokyo's Department of Chemistry more clearly than anything else.

Chapter 6

Making Use of a Pedagogical Space for Pure Chemistry

I [Sakurai] give a course of lectures on Theoretical and Physical Chemistry for 3 hours a week, and another on Higher Organic Chemistry for 3–5 hours a week, both during the whole session, Dr Divers lecturing on descriptive Inorganic Chemistry for 3 hours a week. [In the laboratory,] the students go through preliminary work in analysis and organic preparations, and afterward devote their time to practical work in Physical Chemistry. They also work in the Physical laboratory and attend lectures on Higher Physics and Mathematics. Once a week, we hold Journal meetings, when papers in various chemical journals are read and criticised. It does students good by giving them more occasions for reading original articles, and also for improving their knowledge of foreign literature.

(Sakurai Jōji to Wilhelm Ostwald, June 14, 1898)[1]

Any study of a pedagogical space would not be complete without looking at what actually occurred in it. Space affects but does not predetermine pedagogy. This depends on users. Sakurai Jōji, a pioneer of physical chemistry in Japan with strong penchant for theories, and Edward Divers, an adept experimentalist in inorganic chemistry, had distinct pedagogical and research styles. This is what Sakurai suggested to physical chemist Wilhelm Ostwald, one of the few German colleagues he was friends with. I will look at how differently these two senior professors at Tokyo's Department of Chemistry made use of the same space analyzed in chapter 5 to deliver lectures, supervise advanced students' work, conduct their own research, and connect with students.[2]

On the other side of this equation are the students. This chapter examines how the students, most notably Ikeda Kikunae (physical chemistry), Haga Tamemasa (inorganic chemistry), and Majima Toshiyuki (organic chemistry) responded to Sakurai's and Divers's teaching both as followers *and* dissidents. Based on my spatial analysis, I suggested how fixed pedagogical spaces and a growing number of younger full professors gave rise to an intergenerational conflict in the faculty at the turn of the century, which eventually gave rise to the transformation of the order of chemical education that had prevailed in Meiji Japan (chapter 5). Students' response to teaching is key to understanding its dynamics.

Curriculum Development

The curricula of the Department of Pure Chemistry at the former Tokyo University (see chapter 4, table 4.2) and its successor, the Department of Chemistry of the new Imperial University (table 6.1), show certain continuities. An emphasis on the physicalist approach to chemistry was an important aspect of this continuity. Mathematics and physics also continued to play an important role. Sakurai's course on chemical philosophy developed into his courses on the history of chemical theory and on theoretical and physical chemistry. Another course was added along these lines, i.e., on the use of spectroscopy in chemical analysis to identify elements and to investigate the structures of organic compounds.[3] These courses reflected Sakurai and Williamson's vision of chemical education.

Table 6.1 Curriculum of the Department of Chemistry, College of Science, Imperial University, 1888–1889

	First year	Second	Third
Calculus	Miwa (1,2,3)		
Elementary Dynamics	Knott (1,2,3)		
Higher Physics	Yamagawa (2,3)	Yamagawa (1,2)	
Physical Laboratory		Yamagawa (1,2,3)	
Physiological Chem./Lab	Divers (1)		
Inorganic Chemistry	Divers (1,2,3)	Divers (1,2,3)	
Organic Chemistry		Sakurai (1,2,3)	
History of Chemical Theory			Sakurai (1)
Theoretical/Physical Chemistry			Sakurai (2,3)
Chemical Optics/ Laboratory			Yoshida (2)
Chemical Laboratory	Divers	Sakurai	Sakurai
German	Iijima (1,2,3)	Iijima (1,2,3)	

Note: Numbers in parenthesis shows the terms (3 terms per year) in which each course opened.
Source: Kikuchi 2000; *Jpn. Calendar Imp. Univ.; TDn*, vol. 6.

What disappeared from the curriculum is equally important: The department discarded the application-oriented courses inherited from Atkinson. First, courses on qualitative and quantitative analysis were replaced by a course designated "chemical laboratory." This does not mean the disappearance of analytical training from students' laboratory work at the Department of Chemistry altogether, but it signaled a reduced role of analysis. This shift of focus is understandable if we remember that analytical training at both the Tokyo Kaisei Gakkō and Tokyo University was strongly oriented toward practical applications for industrial, commercial, and public purposes. In contrast, chemical analysis would continue to play a pivotal role in laboratory training in the Department of Applied Chemistry (chapter 7).[4]

In line with this change, application-oriented subjects such as chemical technology, metallurgy, drawing, and assaying were all removed from the curriculum despite the presence of Divers, who might have been expected to teach chemical technology or applied chemistry as he had done at the Imperial College of Engineering, Tokyo. It was in 1898–1899, just before Divers left the department in 1899, that a course of applied chemistry delivered by a professor from the Department of Applied Chemistry was reintroduced to the department's curriculum. In short, it is very likely that Sakurai *and* Divers joined in removing technological aspects from the chemical education of their department.

Yet Sakurai did try to differentiate his and Divers's methods of teaching in his letter to Ostwald. The keyword is "descriptive," which Sakurai used to describe Divers's approach in contrast to his own "theoretical" treatment of chemistry. The question is whether he really could mean anything by using the term "descriptive" for a subject like chemistry that is at least in part inherently descriptive.

Sakurai in the Classroom

Sakurai took his job of lecturing in the classroom seriously, and he expected students to do so as well. A former student recalled that: "If a student did not come to Sakurai's class in time, Sakurai used to wait until the student appeared or send another student to his lodgings to bring him to the classroom."[5] This comment has sometimes been translated differently to suggest that Sakurai publicly reprimanded students for unsatisfactory attendance.[6] However, rather than signifying a "reprimand," such incidents demonstrated Sakurai's view of the importance of lectures for students' learning. This episode also shows the dynamics possible in small classes. Until 1900, the average number of undergraduate students in the department was 11 (chapter 5, table 5.1). It is doubtful whether he would have taken such an action in a larger class of, say, 50 students.

Sakurai acquired his basic attitudes as a teacher from Williamson's lectures, which he admired greatly. According to Sakurai, "[as] a lecturer, Dr. W[illiamson] was always most clear, precise and impressive, and I always took such a delight in hearing him," and Sakurai made explicit that he "acquired [his] habit of endeavouring at clarity and precision from attending [Williamson's] lectures" (see chapter 3). If what Sakurai learned from his mentor mostly came from classroom lectures, it is hardly surprising that Sakurai saw classrooms as his principal contact zone with students, and that's how he designed his pedagogical space (chapter 5). He prepared his lectures thoroughly, selecting "the most trustworthy facts from recent discoveries reported in periodicals mainly published by [specialist] societies

Table 6.2 Thematic annotated catalogue of Sakurai's notebooks

Syllabuses and examination papers

1. "Sketches of Lectures on Organic Chemistry and Physical and Theoretical Chemistry"

Lecture notes for the course of organic chemistry

2. "Organic Chemistry Vol. II"

   Methane and its allied bodies (incl. organometallic compounds), Ethane and its derivatives

3. "Organic Chemistry Vol. III"

   Ethane and its allied bodies (continued, incl. organometallic compounds), Acetyl compounds, Propyl Compounds, Butyl compounds, Amyl or Pentyl Groups, Hexyl groups

4. "Organic Chemistry Vol. V"

   Glycerine and its allied bodies, The 'Carbohydrates,' Urea, Uric acid, and allied bodies

5. "Organic Chemistry Vol. VIII"

   Naphthalene and its allied bodies, Anthracene and allied bodies (incl alizarine), Alkaloids

Lecture notes for the course of theoretical and physical chemistry

6. "Lectures on Chemical Philosophy"

   Historical development of chemistry, as discussed in chapter 4 of this book

7. Untitled Notebook 1

   Covering the determination of molecular and atomic weights, phase transition, kinetic theory of heat, the periodic law, and the relation between physical properties (melting and boiling points, optical properties, such as refractive and rotational power and thermal properties) of substances and their constitutions

8. "Thermochemistry"

   An excerpt in English from Berthelot 1879

9. Untitled Notebook 2

   Discussing chemical affinity, theory of solutions, and electrochemistry of Ostwald's school of physical chemistry [See Kikuchi 2000: 221 for a synopsis]

Source: Sakurai Jōji Papers: C (Lecture Notes)

in Europe and America" rather than using a particular textbook.[7] He thus acquainted undergraduates with the frontiers of chemical research. Lecture notes in the Sakurai Jōji Papers confirm that this was his approach.[8]

Throughout his professorship, Sakurai was in charge of two courses in the Department of Chemistry: organic chemistry and theoretical and physical chemistry. His extant lecture notes are far from complete, but an uncataloged notebook entitled "Sketches of Lectures on Organic Chemistry and Physical and Theoretical Chemistry" helps us understand the overall structure of Sakurai's lectures.[9] It includes several versions of syllabuses and examination papers for both organic and theoretical and physical chemistry and shows the development of Sakurai's lectures until the decade leading to his retirement in 1919. Table 6.2 summarizes the principal features of Sakurai's lectures.

Contents, Structure, and Teaching Methodology

At the center of Sakurai's strategy was the notion of "constitution," which he treated both as a topic in itself and as a heuristic for structuring his subject matter. He introduced his organic chemistry course by describing how to purify and analyze organic compounds, the different types of formulas (empirical, molecular, rational, constitutional, and stereochemical) and the classification of organic compounds. This served as an outline of how his lectures would proceed.

Sakurai's treatment of organic compounds was itself descriptive: For each substance, its mode of formation, preparation, properties, principal derivatives, and reactions were explained. Its constitution was then given the role of unifying these experimental facts. In examinations students were required to show what they remembered and also to use and organize experimental data to discuss the constitution of substances, emphasizing their stereochemical orientation. Though Sakurai selected "the most trustworthy facts" as the subject matter for his lectures, Sakurai left some questions of constitution open, especially in the complicated field of aromatic compounds. His strategy when dealing with such complexity was to both present and criticize contemporary positions. This had the pedagogical function of making students think about what their teacher told them rather than being merely passively receptive.[10]

The organic compounds he selected for his lectures were presented according to increasing complexity in the constitution of the hydrocarbon from which they were derived. His lectures dealt with the whole subject by means of what might be called "a web of derivatives" of core hydrocarbons. Sakurai moved from the simplest methane to higher, branched paraffins, from paraffins through olefins to the acetylene series, and from open-chain through cyclic to aromatic compounds. He concluded the course with highly complex alkaloids. Unlike the format of the lectures of his mentor Williamson discussed in chapter 3, this organization had the disadvantage of referring to compounds with the same functional groups (e.g., alcohols, carboxylic acids, aldehydes, and amines) several times in different contexts; however, this approach gave a coherent structure to his lectures.

Sakurai's statement that he compiled his lectures from original papers without any particular textbook does not preclude preexistent models for the way Sakurai structured his lectures. Williamson's lectures were apparently not the model. Although Williamson used graphic structural formulas in his lectures on organic chemistry, he never used the structural theory as an organizing principle. A likely model for Sakurai's lectures was *The Chemistry of the Hydrocarbons and their Derivatives, or Organic Chemistry*, the third volume of Roscoe and Schorlemmer's multivolume *Treatise on Chemistry*, which his erstwhile colleague Kuhara cited publicly (chapter 4). Though the volume's contents did not cover alkaloids and did not exactly correspond to contents of Sakurai's lectures, its title clearly indicates a similar organizing principle. Indeed, Sakurai mentioned Roscoe and Schorlemmer's *Treatise* in his syllabus as early as the early 1880s, albeit earlier volumes on inorganic chemistry.[11] If this assumption is correct, then rather than rejecting Kuhara's approach to chemistry teaching out of hand, Sakurai integrated it into his own teaching just as he did with the *Zasshi-kai* (chapter 5).

Constitution was also the principal focus of his lectures on theoretical and physical chemistry, in which he frequently mentioned the constitution of organic compounds in conjunction with their physical properties, such as melting and boiling points, refraction, optical and magnetic rotatory powers, and electrolysis.[12] Comparison between measured thermochemical data and data calculated from molecular formulas of organic compounds in homologous series clearly was prompted by the same motive of relating physical properties to constitution.[13] Sakurai introduced recent developments in physical chemistry to Japan under the three headings of chemical affinity, theory of solutions, and electrochemistry, but he did not do so simply for the sake of novelty. Sakurai's stress on the importance of physics for chemistry came from his time studying with Williamson (chapters 3 and 4). For Sakurai, chemistry was "the science of studying the changes caused by the vibration and motion of atoms." The discussion of the relationship between constitution and physical properties would help elucidate this.

He made a clearer statement on the general educational value of chemistry in his letter to Ostwald dated June 14, 1898:

> I am, under the auspices of the Minister of Education, going to give a course of lectures on the elements of modern chemistry to the teachers of all the middle and normal schools.... I hope of convincing [sic] my hearers [sic] that the educational value of chemistry will be greatly increased by treating it not merely as an empirical and descriptive science, but also as a deductive science. Natural history, on the one hand, and Mathematics, on the other, have their own educational value, but only their own. Chemistry of the present day combines the merits of the two, and it is upon [sic] this point that I intend to draw a special attention of my audience.[14]

The context of this quotation and especially his use of the term "modern chemistry" in other parts of this letter show that Sakurai referred to the general educational value of physical chemistry. However, as the above discussion shows, this statement would apply equally to Sakurai's treatment of organic chemistry in his university lectures. Sakurai's lectures were filled with empirical descriptions, but by organizing facts according to the concept of constitution and by making students think about this, Sakurai wanted to add logic to organic chemistry and to enhance the educational value of organic chemistry by "treating it not merely as an empirical and descriptive science, but also as a deductive science."

Similarly, in his popular articles of the 1880s and 1890s Sakurai adopted Williamson's key arguments that constitute the liberal science model. Above all, Sakurai agreed with Williamson's argument about the wider educational value of science, asserting later in life that it taught "the scientific method" (*rigaku teki hōhō*) or "the way to acquire well-founded knowledge," which is relevant to people with a wide range of career prospects and therefore should be an essential part of national education.[15] Sakurai's extramural activities as a key figure in national science teaching and the fact that many of his students pursued careers as teachers meant that he had good reason to be deeply interested in how science was taught in national education in general. This also means that Sakurai's views on how to teach chemistry exerted nationwide influence, as I discuss later in this chapter.

Students' Reactions

Most of the available recollections of Sakurai's students are from those who later became chemistry professors at universities and who thus may not represent the reactions of the whole student group of the Department of Chemistry. Besides, students' recollections are subjective by their very nature. However, such recollections do provide a glimpse of what occurred in the contact zones between teachers and students.

It is clear that Sakurai communicated well in the classroom. His impact was greatest on those students who later became the first generation of Japanese physical chemists.[16] However, he also influenced the next generation of organic and inorganic chemists. For example, Majima Toshiyuki (or Rikō), one of the most productive organic chemists from Tokyo's Department of Chemistry at the turn of the century, studied there between 1896 and 1899.[17] He then worked there as an assistant (and research student) and assistant professor until 1907 and later at the newly established Tōhoku Imperial University (*Tōhoku Teikoku Daigaku*) in Sendai from 1911. Therefore, he was in a position to appraise Sakurai's teaching dispassionately. His image of Sakurai was mixed. On the one hand, he asserted that "Professor Sakurai's lectures on physical and organic chemistry were both reasoned and systematic, and I found them very interesting and useful." He also "admired how Professor Sakurai did everything in an orderly manner." On the other hand,

Majima was less happy with Sakurai's supervision of students' experimental work (chapter 5).

Shibata Yūji, who studied in the Department of Chemistry a bit later, between 1904 and 1907, and joined its teaching staff in 1913, also left recollections of Sakurai's teaching. Shibata was enthusiastic about Sakurai's lectures: "Professor Sakurai was a clear and fluent speaker, and we could make a really organized notebook if we recorded his lectures as though they were dictation."[18] He observed further:

> The lectures of Professor Sakurai were very good. I had the desire to specialize in organic chemistry because I was fascinated by Professor [Sakurai]'s lectures. They were not an exhaustive survey of organic substances, but a very systematic treatment of organic structural theory and stereochemistry. He explained [the] latest researches on structural theory in a systematic way, which was very interesting. I was fascinated by it.[19]

Shibata started his career as an organic chemist, tackling the problem of determining the stereochemical constitution of products from the action of the Grignard reagent upon phthalic acids, which he was first introduced to in Sakurai's lectures.[20] Sakurai's interest in stereochemistry literally "infected" Shibata and provided him with the core question of constitution that drove him throughout his career even after he changed his specialization, on Sakurai's recommendation, to the chemistry of coordination complexes.[21]

Languages and Translation

The question of language is important to understanding how Sakurai communicated with his students in the classroom. An examination of the collection of Sakurai's lecture notes shows that he wrote all chemistry-related notes in English. He used English in making abstracts from books or papers in European languages and also in his own writing on chemistry. Sakurai thus seems to have done all his work as a chemist in English, translating into Japanese only when he needed to prepare Japanese versions of his research papers or to give popular lectures. For university lectures he did not feel that need.

Sakurai indeed gave at least some of his lectures in English. His lectures on chemical philosophy in 1882–1883 at Tokyo University were in English (chapter 4). Ikeda, graduated in 1889, also commented that Sakurai delivered lectures in English.[22] However, we cannot automatically conclude that his lectures were always given in English in later periods because teaching in Japanese became the norm by 1884 throughout Tokyo University and its preparatory schools as the number of foreign teachers decreased. This Japanization continued in the Imperial University.[23]

Collating an excerpt from Sakurai's English-language lecture notes on organic chemistry with the corresponding part of Majima's notes of Sakurai's lectures provides further evidence on matters of language (see appendix). Occasional mistakes aside, Majima skillfully summarized Sakurai's lectures without compromising their contents. One of the most important characteristics of Majima's notes is his use of "creole" or mixed language. Each sentence was constructed in Japanese, but in addition to chemical formulas, most of the names of elements and compounds, technical terms (e.g., hydrolysis, reduction, and derivative) and a few common words such as "heating" and "confirm" were written in English. It is unclear whether it was either Sakurai or his students who devised this creole language for the classroom. However, Sakurai was in a better position to engage in such intellectual acrobatics because he would have known exactly what he wanted to convey in both English and Japanese, whereas his students would have to respond to it. Shibata's comment quoted above, that students felt no difficulty in taking notes in Sakurai's classes, also suggests some kind of linguistic device on Sakurai's part, as students in later years were less well prepared in English.

In any case, whatever his practice, it was the outcome of an interplay in the classroom between Sakurai and his chemistry students. Sakurai had every reason to wish to use English. He had studied chemistry in English and most probably used it for all his intellectual work as a chemist throughout his professorship at Tokyo. Most important of all, his stance as a chemistry professor was that of a Westernizer: he intended a direct transfer of several things from Western countries to Japan, ranging from Williamson's ideal of "pure" science and his physicalist and dynamic view of chemistry to the latest chemical research in Europe and America. On the other hand, few of the chemistry students at the Imperial University in the 1890s were as proficient in English as Sakurai's generation of chemists. They had been trained by Japanese teachers up to the secondary level. As will be seen in the case of Divers's students, the level of proficiency in university students' oral communication in foreign languages also declined dramatically in the 1890s. The use of a mixed language in Sakurai's classroom benefited both parties, making the transfer process smoother. What, then, was Divers's response to this problem? How did he organize his lectures?

Divers in the Classroom

Information on Divers's lectures had to be assembled from various sources because his lecture notes and syllabuses are not available now. Students' recollections are useful to get an image of Divers's lectures as a whole. According to Ōsaka Yūkichi, who studied at the Department of Chemistry between 1889 and 1892, in his lectures Divers "explained the chemistry of each element in detail and did not talk much about theories. So the order of lectures seemed immaterial."[24] Majima also recalled that "in the lectures of Professor Divers the sections on sulfur and nitrogen, which constituted his main research area, were minute, and his explanation of metal elements was rather simplified."[25] Divers likely adopted the "catalog" style to the teaching of inorganic chemistry, proceeding from one element to another, and very likely his treatment of each section was uneven. Students could not sense that the order of lecture material was based on classification or theoretical considerations.

Sakurai's characterization of these lectures as "descriptive" would thus seem to have indicated a lack of organization based on principles. Divers's annual report to the Imperial University for 1887–1888 qualifies this:

> The object of my lectures is to teach inorganic chemistry step-by-step in accordance with students' progress. . . . The attention of first-year students was drawn to the outlines [of general theories and laws], and second-year students have benefited from it in the [exhaustive] studies [of chemical phenomena] from beginning to end.[26]

That is, in his lectures, Divers provided general theoretical considerations as an introduction to a more detailed catalog-like description of chemical phenomena in Divers's lectures. His earlier syllabuses at the Imperial College of Engineering, Tokyo also seemingly followed this pattern.[27] However, despite his positive self-appraisal, it seems doubtful that this way of presentation necessarily worked as Divers intended to with his students at the Imperial University. Students' recollections suggest that only the descriptive part of Divers's lectures left a vivid (and rather negative) impression in their minds, in contrast with their appraisal of the clarity of Sakurai's constitution-based lectures on organic chemistry.

The problem of lecture organization was certainly there, but that of language was another factor in an occasional malfunctioning of communication between Divers and his students. All sources I cite below, especially Divers's examination papers, show that he lectured in English throughout his long tenure at both Imperial College of Engineering and the Imperial University. Ōsaka recalled that around 1890 he found it difficult to take notes in Divers's lectures. He was fortunate enough to be able to borrow a notebook from a classmate who was better attuned, but during the final few years of Divers's tenure his language problem became even more serious. According to Ikeda, after being appointed assistant professor at Tokyo's Department of Chemistry in 1896 he had to attend Divers's lectures with students in order to explain the content to them afterward in Japanese, because no student could understand Divers's lectures. It was a situation reminiscent of Nankō's "sound teachers" in the early 1870s (see Takamatsu's recollection in chapter 3).[28]

The collection of Divers's examination papers in Tokyo's Department of Chemistry indicates that not only Divers's questions, but also the answers of his students were written in English.[29] Students' answers often contained corrections by Divers, and this corroborates Ōsaka's recollection below about tutorials. Divers's examination papers also confirm students' recollections about the "catalog" style of his course coverage and show how he treated applied chemistry in the framework of pure chemistry.

Particularly interesting is his way of presenting experimental facts by avoiding theoretical hypotheses, such as the stereochemical representation of molecular constitution, as much as possible. According to his former student and colleague, Haga:

> [According to Professor Divers,] molecular structural theory is nothing but a kind of picture drawn to recreate chemical reactivity of substances within scholars' brains. . . . He often liked to discuss the question of valency, which was probably due to his interaction with Mr. Kekulé when he was young. According to Professor Divers, valency designates the time a chemical compound undergoes reactions. For example, the tetravalency of carbon means that it takes four reactions to replace all elements attached to carbon with new ones.[30]

Haga's mention of Divers's interaction with Kekulé in this context is telling, as Divers's definition of valency reflects organic structural theory in the 1850s long before it underwent a stereochemical turn in the 1870s and 1880s.[31] It was in this later period that Sakurai trained as a chemist at UCL with Williamson, an early advocate of three-dimensional thinking in organic chemistry, and deeply affected by the emerging subdisciplines of stereochemistry and physical chemistry. The difference between Divers and Sakurai shows the enduring effect that a training received at a young age could have on a scientist's teaching method despite being well-informed of subsequent developments.

Divers eventually earned his students' respect at the Imperial University.[32] He found other ways of communicating with Japanese students to compensate for his linguistic handicap as lecturer. Divers held informal person-to-person tutorials with junior students. According to Ōsaka's recollections, he invited all students in his classes to his office after each semester's examination in order to explain the questions in detail and correct students' answers.[33] Haga testified that the main object of Divers's examinations was not to grade students, but to check whether they had really understood his lectures. He gave model answers to them for comparison with their own examination papers.[34] This method worked with the small classes in the 1880s (with less than ten students), but not with the larger classes he had in the 1890s, occasionally numbering more than twenty students (chapter 5, table 5.1). Hence his problem just before his retirement in 1899.

Divers's invitation of his students to his office was not a breach of the "division of labor" rule set up by Sakurai and Divers because it was an extension of his classroom teaching. But this also shows Divers's flexible use of the pedagogical space due to his language problems, something that students would not expect from Sakurai. Another example of such flexibility is the episode about silver nitrate Ōsaka recalled. One day Haga ordered Ōsaka to collect scrap silver (which was naturally oxidized) from bins throughout the department and to reduce it to make a large lump of pure silver. Ōsaka prepared pure standard silver nitrate solution with this silver (thinly foiled) and then a potassium nitrite solution from this silver nitrate solution. During this operation, however, he mistakenly poured some silver nitrate into his brand-new bench and made a large stain. Divers noticed it and jokingly and laughingly asked Ōsaka: "Are you dyeing your table?"[35] An interesting point about this seemingly ordinary episode is that Ōsaka did his experimental work in a student laboratory for Divers's use in his own separate private laboratory, connecting two parts of the pedagogical space through Haga, the mediator.

Divers's Research Supervision and Research Partnership

As Ōsaka's episode suggests, another way to compensate for Divers's language handicap was research supervision. As I discussed elsewhere, Divers used his experience at the RCC and Queen's College Galway to foster group research by letting the seniors assist the juniors in his department at the Imperial College of Engineering, Tokyo. Divers thereby initiated promising students into original research, guided not by the potential applicability of their topics to industrial purposes but by the quest for "original observation," which often resulted in his publishing joint papers with them.[36] After his move to the Department of Chemistry, Divers did not need to train assistants or seniors anew. With the help of his former student and collaborator, Haga, Divers was able to introduce this practice in his new department straight away.

An analysis of publications and citations of research papers from the Department of Chemistry offers a useful insight into the range of Divers's involvement with the research of advanced and postgraduate students such as Ōsaka, Hada Seihachi, Chikashige Masumi, and Ogawa Masataka.[37] In some instances, Divers acknowledged students' contributions in his papers.[38] In others, Divers granted them joint authorship.[39] There were also instances of students acknowledging Divers's contributions in their papers.[40] At the Imperial College of Engineering, Divers had collaborated as a supervisor only in the form of joint authorship, whereas all three forms of collaborative interaction were taken when he was at the Department of Chemistry, which shows his maturation as a research supervisor. In comparison with Williamson and Atkinson (always allowing students to use their own names as the single author of their works) and Remsen (mostly with his own name as the lead author), Divers adopted a much subtler way of indicating students' contributions to research.

In instances where acknowledgements were given, the degree of teachers' and students' respective contributions was clearly set out. For example, Haga and Divers, in a joint paper they wrote, thanked students for their assistance, but they also made clear that they had retained overall control of the research.[41] In papers where students acknowledged Divers, this tended to take the form of expressions of thanks for "suggesting" topics or starter reagents and for help in writing up papers in English or for "guidance" in carrying out their research, thus suggesting much looser control of Divers over these students' work.

That said, a majority of Divers's collaborations appeared in the form of joint papers with students. The partnership between Divers and Haga undoubtedly functioned as a model for the students. Summing up their work on a series of some ten papers, Divers himself made clear that: (1) the choice of research topics and the planning of investigations was done solely by Divers; (2) during the course of the investigation both Haga and Divers came up with new ideas, some of which became important discoveries; and (3) that the experimental work was largely executed by Haga.[42] As these statements constituted part of the examination for Haga's doctor of science degree, there is little possibility that Divers understated Haga's contributions there.

What stands out here is Divers's skill as a research project coordinator. Following an eye injury at the Imperial College of Engineering, which made him almost blind in his right eye, Divers apparently did less work at the bench himself than previously, but in the hands of his students and assistants his output continued to be highly experimental.[43] Divers also collaborated with peers from other colleges within the Imperial University. During his researches with Haga on amidosulfonic acid, he asked Sakurai, Oscar Loew (German physiological chemist at Tokyo's College of Agriculture), and Tahakashi Juntarō (Japanese pharmacologist at the College of Medicine) to do physico-chemical, physiological, and pharmacological investigations of this acid and secured the publication of the results in the *Journal of the Chemical Society: Transactions*.[44] In Sakurai's case, this request stimulated a new line of research. Considering the institutional independence of each college within the Imperial University, this example illustrates Divers's ability to coordinate interdisciplinary research.

I have thus far shed light largely on the positive aspects of Divers's research supervision. However, there is no perfect way for training scientists. Some historians suspected that Divers's strong leadership in his joint researches might have suppressed the originality or creativity of Japanese collaborators, an important question worthy of close scrutiny.[45]

On the one hand, during his professorship at the Imperial College of Engineering Divers formulated a view of the educational value of laboratory training as essential for developing "powers of close observation of the phenomena," "a capacity for making original observation," and "the ingenuity and perseverance necessary to give fruit to their [students'] observations."[46] If we trust his words, we have to say that he did not treat his collaborators as mere laborers, but encouraged them to propose new ideas, which occasionally resulted in discoveries of new experimental facts. In this sense, Divers's supervision can hardly be considered to have suppressed students' creativity. On the other hand, Divers took strong control of students' work at least by assigning their research topics. This is doubtlessly one of the essential features of research schools, and Divers's joint researches with students had a better chance of success as a result. However, this entailed the potential risk of depriving students of any opportunity to learn to devise research topics on their own from scratch.

Divers's former student, Ogawa Masataka is a case in point. After joint research with Divers at Tokyo, Ogawa went to England for overseas study between 1904 and 1906. Ogawa chose to work at UCL with William Ramsay (Williamson's successor from 1888 on) who assigned Ogawa the research topic of isolating new elements from a sample of thorianite. This assignment largely determined Ogawa's scientific career as a researcher on "Nipponium" after he returned to Japan in 1906 and took up the professorship of inorganic chemistry at Tōhoku Imperial University in Sendai in 1911.[47]

Divers's impact on Ogawa revealed itself in two ways. On the one hand, Ogawa was proud of the solid analytical technique, the "orthodoxy" of chemical research, which he learned from Divers.[48] Ogawa's work on "Nipponium" mainly comprised the painstaking separation of a new element from the sample by classical analytical methods, the only exception being the spectroscopic analysis Ramsay provided for identifying Ogawa's "new element."[49] Here, Ogawa's analytical skill and his knack for "original observation" trained by Divers counted for much. On the other hand, Ogawa never looked for a research topic on his own, did not take up a new research topic fundamentally different from "Nipponium" after he returned to Japan, and in the early years of his professorship he required students and assistants to be involved in his research project on "Nipponium," which largely frustrated them.[50] As the stake in his "Nipponium" research was so high, the above discussion might sound too harsh on Ogawa, but it points out the pros and cons of Divers's supervision of students' research at the Imperial University through Ogawa's example. Now is the time to fully appraise Sakurai's alternative approach to research supervision.

Sakurai's Research Supervision and "Individualism"

In chapter 5 and earlier in this chapter, I cited Majima's mixed feeling about Sakurai's supervision that "while he welcomed the freedom Sakurai gave to him, it made him feel a little bit insecure." That quote points to both the question of, and the difficulty in, assessing Sakurai as a research supervisor. Did Sakurai give freedom to students under his supervision for some purpose, or did he just leave them alone?

To be sure, Sakurai's involvement in students' work looks totally different than that of Divers. Sakurai did not acknowledge any assistance from students or assistant professors in his own researches, and he never published his papers with joint authorship. Furthermore, his students who wrote papers on topics in organic chemistry were largely silent about Sakurai's supervision. For example, Majima's largely negative comment about Sakurai as a supervisor is corroborated by a perusal of Majima's own research papers published during his assistantship and assistant professorship at Tokyo, none of which mentioned Sakurai.

Students' work in physical chemistry, in contrast, points to a different story. Ikeda, who studied physical chemistry under Sakurai for one year after his graduation in 1889, published a paper on the relationship between capillary attraction and chemical composition based on R. Schiff's data, in which he mentioned Sakurai's supervision.

> In conclusion I have to return my best thanks to Professor J. Sakurai, for the great interest which he has taken in my work and for his valuable suggestions.[51]

His wording, "the great interest he has taken in my work," is a level of appreciation that students working with Divers never expressed. It would seem that Ikeda chose his research topic for himself: otherwise his mention of Sakurai's "great interest" would have made little sense. He did not explain Sakurai's "valuable suggestions" in detail, but Sakurai's lectures in theoretical and physical chemistry mentioned Schiff's work on surface tension and, more important, Sakurai later inserted "Ikeda's expression [his research result expressed in equations]" into this part of his syllabus.[52]

The following is a plausible scenario: Ikeda first heard of Schiff's research on surface tension in Sakurai's lectures on theoretical and physical chemistry and proposed the extension of Schiff's research as his research topic when he was enrolled as a postgraduate student.[53] Sakurai was clearly interested in the project and made suggestions during the experimental phase and gave Ikeda language support when the latter was writing up the English manuscript.[54] Sakurai may possibly have arranged the publication of Ikeda's paper in the college journal and then completed the circle by integrating Ikeda's results into his own lectures.

Indeed, Ikeda's relationship with Sakurai in the role of supervisor did not end in 1890 when it officially terminated. Ikeda subsequently acknowledged Sakurai's "valuable advice" on a later project on the effect of solvents on solutes' reaction rates.[55] Sakurai, again closing the loop, used part of Ikeda's research in a lecture demonstration in the Ministry of Education's summer course for schoolteachers (*Monbushō Kaki Kyōin Kōshūkai*) in 1898.[56]

Following the example of Williamson, Sakurai took a more "individualistic" approach to research supervision than Divers and regarded students' researches as their own individual enterprises. Sakurai did not adopt joint authorship in students' work and left no evidence indicating that he assigned research topics to advanced or postgraduate students or ordered students to assist in his own experiments. His role as supervisor was restricted to that of occasional advisor during students' experiments and the writing up of papers, especially in English. The journal published by the College of Science from 1887 on had the policy of accepting only articles written in Western languages. Though, as noted above, students' proficiency in foreign language declined in the 1890s, the College of Science as an institution encouraged professors and students to publish papers in Western languages. Sakurai's language support was important in this context.

Sakurai's Research Activities: Research as Part of Teaching

Sakurai's colleagues and students, as well as subsequent historians, agreed that the most important achievement of Sakurai's research activities was his improvement of Beckmann's method of molecular weight measurement using the rise in boiling points of solutions.[57] Sakurai, who took his teaching activities seriously, had indeed a good reason to take this topic. The measurement of boiling points of solutions and their use in determining molecular weights played a pivotal part in Sakurai's lectures on organic chemistry and also in the lectures on theoretical and physical chemistry. Likewise, the technique occupied a fundamental place in students' experiments in analysis and organic synthesis because the determination of molecular weights was a prerequisite to fixing molecular formulas. To understand how Sakurai's teaching and research were interwoven with each other, it is necessary to look at this research in some detail.

Sakurai's approach to this problem was that of experimental physics and included precision measurement, in which he was well-versed and trained through Williamson's lectures on chemistry and through the lectures and experimental sessions in physics by George Carey Foster and Oliver Lodge at UCL. In his first paper on this subject published in the *Journal of the Chemical Society: Transactions* in 1892, Sakurai first raised the question that would be central to all his work in this area—whether the temperature of steam rising from a boiling salt solution was the same as that of the solution.[58] According to Sakurai, this had not been settled by the previous work of physicists, such as Faraday, Rudberg, Magnus, and Müller. Sakurai identified the fluctuation of temperatures by irregular boiling and bumping as the main difficulty in the measurement, and he devised a method for introducing "steam into the boiling solution from without," so that "evaporation and condensation of steam in the solution can be so readily and exactly counterbalanced" and "its boiling temperature may be maintained constant for any length [of time]," using, implicitly, Williamson's concept of dynamic equilibrium.[59]

In his next paper published in the same journal, Sakurai used this technique of introducing solvent vapor into a solution to effect an improvement on Beckmann's method of measuring molecular weights by the rise in boiling points of solutions (see figure 6.1).[60] Sakurai pointed out that Beckmann's apparatus needed great care to prevent the flask from cracking and claimed that his own apparatus was "exceedingly simple and can be set up by any one with materials commonly found in all laboratories."[61] He thought it simple enough to be used by inexperienced researchers, even students. Though Sakurai did not establish a research school, he thus demonstrated one of the characteristics of research leadership in establishing a technique or an apparatus that could be readily used by inexperienced researchers, just like Liebig's *kaliapparat*.[62]

Sakurai's key research reflected the materiality of his milieu at Tokyo at least in two ways. First, Tokyo's Department of Chemistry at that time still relied for its glassware on imports from Britain, and the repair of a more

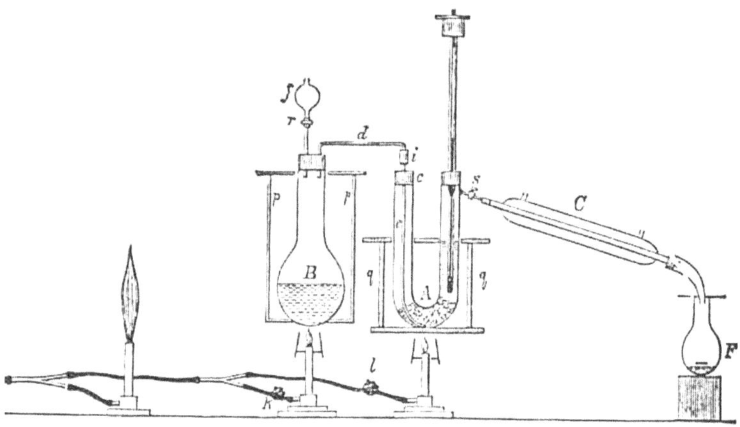

Figure 6.1. Sakurai's apparatus to measure the boiling point of a solution for the determination of the solute's molecular weight. Note that the burner with a bigger flame functioned as a gas regulator.
Source: Sakurai 1892b

sophisticated apparatus such as Beckmann's exceeded the skill of glassblowers who were mostly professors, like Ikeda, and students. Sakurai had a good reason to devise an "exceedingly simple" alternative. Second, the existence of a "gas regulator" (the gas burner with a larger flame on the left) to combat fluctuation in gas pressure in his apparatus suggests unstable gas supply for his laboratories; that supply was not part of urban infrastructure in Tokyo but came from the university's own gas station. In spite of these local origins, Sakurai's idea was soon taken up and improved by other chemists, such as Landsberger, Walker, and Ikeda, and was used widely to measure molecular weights until the advent of the mass spectrometer in the 1950s.[63]

The other research area Sakurai tackled was the relationship between physical properties and the molecular constitution of chemical substances, a frequently visited lecture topic. Electrical conductivity was a particular focus and occupied a large proportion of Sakurai's lectures on theoretical and physical chemistry. In his two papers in the *Proceedings of the Chemical Society* in 1895, Sakurai argued that glycocine (NH_2CH_2COOH, glycine or glycocol in today's nomenclature, Sakurai used both glycocine and glycocol) forms "an internal ammonium salt" and should be represented by a closed (cyclic) formula instead of an ordinary open-chain one on the basis of low electrical conductivity and a low rate of ionization of glycocine.[64] Divers's request to determine the molecular conductivity of amidosulfonic acid (NH_2SO_3H) gave Sakurai an opportunity to revisit this question. He compared the reverse influences of the amino group ($-NH_2$) on the strength of organic and sulfurous acid (organic acids becoming weaker with the presence of an amino group and sulfurous acid stronger) and reasserted his "internal ammonium salt" theory of glycocine on the basis that organic acids with an amino group are stereochemically more likely to form "internal salts" than amidosulfonic acid.[65]

That Sakurai was involved in no joint publications does not mean that he worked in isolation. I have already mentioned that Sakurai's researches on molecular conductivity of amidosulfonic acid formed part of Divers's larger research project on this substance. In a key paper on the determination of the temperature of steam from boiling salt solutions, he mentioned Divers's "many valuable criticisms and suggestions from time to time while this investigation was in progress" as well as his physics colleagues, Cargill G. Knott and Yamagawa Kenjirō, who had taken an interest in Sakurai's work.[66] Equally interesting is that Sakurai's students later tried on Beckmann, Sakurai, Landsberger, and Ikeda's methods and compared their results.[67] This suggests the students' role in making Sakurai's technique more practical and stable, similar to what Catherine Jackson argued for Liebig's *kaliapparat*.[68]

Most important, Sakurai's research activities were closely interwoven with his lectures, which makes it difficult to distinguish between them. As we have seen, Sakurai's lectures were based on current international research and incorporated excerpts from recent papers. Not only did Sakurai's own research activities come from his lecturing, but the results of his research were also integrated into his lectures and inspired students such as Ikeda, some of whose results were again incorporated into Sakurai's lectures.

Shibata's obituary of Sakurai likewise points to the power of his lectures to influence his students' research trajectories:

> Of course, [Sakurai] sensei's lectures mentioned his famous research on molecular weight measurement using the rise in boiling points of solutions. But I also remember listening with great interest to his talk [in his lectures] on his confirmation of the cyclic constitution of glycocol by the minute fact of its electrical conductivity and taking a quick note of it. This achievement should be regarded as a pioneering work on auxiliary valency and inner complex salts and has great significance in modern chemistry.[69]

Importantly Shibata remarked on Sakurai's emphasis on the importance of physico-chemical data to determine molecular constitution, how interesting that was to a young student like him, and how he understood the significance of Sakurai's work in the context of his own future specialty, coordination chemistry, where physico-chemical data such as spectroscopy played a key role. Shibata also provided indirect evidence about Sakurai's integration of his experimental technique into the laboratory teaching at Tokyo. He astonished fellow researchers in Leipzig with his demonstration of how to determine molecular weights by boiling point measurements, which he had learned in Tokyo;[70] this was undoubtedly Sakurai's or Ikeda's improved method.

In short, Sakurai's research-based teaching inspired students to engage with current international research frontiers and emphasized that physico-chemical experimentation was indispensable to compete internationally. Even students of organic chemistry who were less engaged in physical chemistry found Sakurai's integration of teaching and research inspiring. Majima, usually critical of Sakurai, bound his notes on aromatic compounds, then the most advanced and complicated part of Sakurai's lectures on organic chemistry, and kept them throughout his life.[71] As I mentioned, Sakurai compiled them from recent original papers published in Europe and America, introducing current discussions about the constitution of complicated aromatic compounds. Majima's technique of what he called "the research of great researches," i.e., a thorough literature review of the work of major organic chemists in Germany, reflected the best of what he learned from Sakurai.[72]

Conclusions: Ikeda as the Follower and Dissident of the Department Regime

This chapter is very much about how students reacted to Sakurai and Divers's pedagogical styles as both followers and dissidents. Ikeda's later career as full professor of the department exemplifies this dynamic and merits some discussion here. It may sound paradoxical, but Ikeda's reassertion of independence from Sakurai, especially after his return from overseas study in Leipzig with Ostwald between 1899 and 1901, is indicative of Sakurai's "individualistic" approach. This was evident on a couple of fronts. I already mentioned Ikada's laboratory design that deviated from Sakurai's. Ikeda's commitment to Ostwald's energeticism contrasts clearly with Sakurai's atomistic worldview.[73] The most conspicuous and boundary-crossing, though, is Ikeda's most celebrated research on "new seasoning," i.e., monosodium glutamate (MSG).[74] It is the kind of research that we would not expect to be undertaken in Sakurai's laboratory, as is clear from the above review of his research projects.

According to an autobiographical sketch written much later in his life, Ikeda's interest in chemical technology predated his attendance of Tokyo's Department of Chemistry between 1885 and 1889, and he tried to incorporate the explanation of as many manufacturing processes as possible into his lectures on physical chemistry at Tokyo.[75] However, it was right after his return from Germany that he made his position public for the first time in 1902 in *Tokyo kagaku kaishi*, the organ of the Tokyo Chemical Society.[76] An English translation of the title of his Japanese article is "On the relationship between preparative chemistry and physical chemistry," but it does not really convey his true intention.

First, Ikeda used German *präparative Chemie* instead of the English term "preparative chemistry" in his article, revealing the impact of his experience in Germany. As he pointed out, most of organic and inorganic chemistry should really be called *präparative Chemie* because chemists, especially German organic chemists, had long used preparation for analytical purposes to guess and ascertain the structure of compounds (see chapter 3, the discussion of Kuhara's graduate thesis at Tokyo).[77] Ikeda's argument was then little different from a typical rhetoric employed by physical chemists to present their field to the wider chemical community. By using physico-chemical theories or principles on reaction rates and chemical equilibria, organic and inorganic chemists could better predict and control the yields of the chemical species they were preparing.[78] More important, Ikeda cleverly adopted "goods-producing chemistry (*seihin kagaku*)" as the Japanese translation of *präparative Chemie* to highlight the relevance of his argument to chemical technology for readers without any detailed discussion.

His research on MSG was partly his way of putting his 1902 article into practice. His key recognition that monosodium glutamate, not glutamic acid, caused *umami*, the fifth taste sensation previously unrecognized, would have required a rudimentary understanding of the ionic theory, which Ikeda certainly had studied, unlike his peers in, say, organic chemistry.[79] His reasoning on how differently the dilution of MSG (weak electrolyte) and common salt (strong electrolyte) affects the taste and on the effect of vinegar on the taste of MSG is a classic example of the application of Arrhenius' theory of dissociation and Le Châtelier's law.[80]

Perhaps, however, the most intriguing aspect for today's readers is at the end of Ikeda's paper. He disclosed that he had already patented the production method of MSG as a new seasoning in July 1908, well ahead of the publication of this paper in 1909. He added that the chemical manufacturer Suzuki Saburōsuke had already started the production and sale of MSG, and that it would revolutionize the manufacturing method of soy sauce as well because part of its taste would undoubtedly be attributable to MSG.[81] It would be difficult for us not to detect a shrewd "business sense" encouraged by Nakazawa Iwata at Tokyo's Department of *Applied* Chemistry, but not so at Sakurai and Ikeda's department (chapter 7). The patenting was a perfectly logical and natural behavior for Ikeda, if not for Sakurai. Ikeda seamlessly and self-consciously moved between cutting-edge science and industry and between knowledge production and moneymaking and defied the Meiji distinction between pure and applied science.

Ikeda was a shrewd observer of the management of Tokyo's Department of Chemistry as well. He obviously knew Sakurai well as his brother-in-law (Ikeda married a sister of Sakurai's wife).[82] But he also closely observed Divers, whose teaching he once supported as assistant professor. While praising Divers's achievements, Ikeda was critical of his way of mentoring his students, especially Haga, a way Ikeda thought deprived him of the opportunity to become an independent researcher, a theme taken up also by later historians.[83] Thus, Ikeda implicitly asserted the value of the freedom he enjoyed under Sakurai. Ikeda also correctly pointed out that the lack of connection between his department and industry was closely related to the employment pattern of graduates.[84] Up until the early 1900s, most graduates of Sakurai's department pursued teaching careers, not only at imperial universities but also at secondary schools, higher normal schools (*kōtō shihan gakkō*, schools for training teachers of secondary schools) and even at middle schools (*chūgakkō*, the secondary-level schools that prepared students for higher schools). Movement of teachers between these institutions for tertiary and secondary education was high during the Meiji period.[85]

Sakurai himself did not leave any official statement about his view of the social functions of his department or the College of Science unlike his mentor Williamson in the *UCL Calendars* or in his inaugural address as dean of Science. However, Watanabe Hiromoto, first president of the Imperial University, stated in his address during the university graduation ceremony in 1887 that the mission of the College of Science was first "to investigate the universal laws and the principles of medicine and engineering to broaden and develop their validities," and second "to cultivate the ideals of young beginners."[86] He had expressed a similar view in his address to the annual meeting of the Tokyo Chemical Society in 1886 (chapter 4). This time he spoke on behalf of the whole professoriate of the university.[87] The second educational function of the college was met additionally by the "Simplified Course" of the College of Science (*Rika Daigaku Kan-i Kōshūka*) in 1889 for training science teachers for secondary schools. This lasted until 1893. Sakurai also taught the chemistry segment of this course throughout its existence.[88]

Outside the university Sakurai came to play a central role in national science education by the 1890s, and, as I discussed, he advocated a system that emphasized the importance of teaching theoretical principles of chemistry and experiments demonstrating these principles at the secondary level to teach "the scientific method." Sakurai occupied several governmental posts that controlled chemistry teaching at the secondary level on the national scale, such as examiner for the state examination qualifying secondary school chemistry teachers.[89] He was also a member of the committee for the screening of government-designated science textbooks for secondary schools and a designer of the national curriculum for chemistry in secondary schools at the turn of the century.[90] Sakurai's above-mentioned involvement in the summer course for schoolteachers of the Ministry of Education was comparable to what Edward Frankland did in the summer course at the RCC for teachers in the classes for Department of Science and Art examinations in the 1870s.[91] This is another sign of Sakurai's central role in national chemistry teaching and of his influence on how chemistry was taught at the secondary level.

The link between Sakurai's network with the educational sector and the careers of his students as teachers was his role of placing graduates in academic and more general educational positions. As a central figure in national science education in Japan and as a teacher of students ideally prepared for the system he advocated, Sakurai was often asked by the Ministry of Education officials and principals of secondary and higher normal schools for suitable candidates. The recollections of his students cited above unequivocally mention Sakurai's power in determining their academic careers. Majima asserted that "I am far inferior to my mentor Sakurai sensei in helping younger [chemists] get suitable positions."[92] The network of the Department of Chemistry with the educational sector of the Meiji society was formed by this role.

Ikeda was familiar with all these mechanisms because he functioned as part of them: He taught chemistry at the Higher Normal School (*Kōtō shihan gakkō*) between 1890 and 1896 before being appointed assistant professor at Tokyo, and he edited several textbooks for secondary schools.[93] His students unequivocally testified to his enthusiasm and skill in teaching.[94] At the same time, he was clearly frustrated by the situation where most graduates took up teaching positions, and there was a lack of connection between the department and industry. Just as Ikeda's frustration as a chemical pedagogue likely motivated his lab design, his MSG research was clearly motivated by his desire to break free from the distinction between pure and applied science characteristic of the Meiji period.[95] Intergenerational conflict had productive effects on the development of the department.

This network-building had its parallel in the Department of Applied Chemistry, due essentially to the character of the imperial university system in general, which saw the role of each college as a training institution (chapter 4). The kinds of manpower each college would produce and the network between college professors and other institutions, such as government agencies, private companies, schools, and hospitals, were determined by such factors as external student loans and graduate careers. Also, the Department of Applied Chemistry had its fair share of intra- and intergenerational conflicts. It is time to move to the Department of Applied Chemistry to close the circle.

Chapter 7

Connecting Applied Chemistry Teaching to Manufacturing

Soak silk overnight in a solution of alum or aluminum chloride that is made slightly basic with the specific gravity of 8 to 12 units (using Twaddell's hydrometer [which was widely used in Britain]) and steep it in a warm 4 percent solution of calcium carbonate or in a solution of sodium silicate (1 unit of specific gravity on Twaddell scale) and rinse it thoroughly with water. Then put the silk into the cold liquid with 15 percent of muddy yellow alizarin and 2 percent of calcium acetate and gradually bring it to the boil for one hour and soak it for another 20 minutes. Rinse it with water and then put the silk in a soap solution mixed with a touch of stannous chloride and warm it. If the color is not bright enough, warm the silk in the soap solution again and rinse it in water. Finally, soak the silk in water containing acetic acid or tartaric acid and wring and dry it. You will get a brighter red color if you add a small amount of turkey-red oil.

(Takamatsu Toyokichi on red dyeing silk with alizarin, 1895)[1]

Changes in the academic landscape of chemistry in Japan in March 1886 created two departments with the common origin in Atkinson's teaching program at Tokyo University: the Department of Chemistry, which has been examined in chapters 5 and 6, and the Department of Applied Chemistry. An ideal way to understand how pedagogically different these departments were would be to enter classrooms in both departments and listen carefully to lectures. If this is difficult or impossible, the second best method would be to read a textbook written by one of their professors, such as the one quoted at the beginning of this chapter. You will soon find that this is no ordinary textbook of chemistry. Indeed, the Department of Applied Chemistry offered no ordinary chemistry teaching. The focus of this chapter is to understand how this came about.

The College of Engineering of the Imperial University was established by a merger of the Faculty of Industrial Arts of Tokyo University and the Imperial College of Engineering, Tokyo (see chapter 4). The Department of Applied Chemistry was one of seven departments of the college. They were temporarily housed in the building of the former Imperial College of Engineering in Toranomon before a building for the College of Engineering was completed on the Hongō campus in July 1888.

Two events at quite an early stage affected the development of the department. The first is the transfer of Edward Divers, principal and professor of practical chemistry at the Imperial College of Engineering, to the College of Science, not Engineering. Takamatsu Toyokichi and Matsui Naokichi from the former Tokyo University were predominant as full professors, while Divers's former students and assistants at the Imperial College of Engineering, Kawakita Michitada and Nakamura Sadakichi, occupied a subordinate position as assistant professors.[2] This is exactly the opposite of what happened to many departments of the College of Engineering, exemplifying the necessity of discipline-specific scrutiny of the history of technical education.[3] We can only speculate on what would have happened if Divers would have stayed in the College of Engineering, but the effect of his move on the balance of power in the faculty of the Department of Applied Chemistry seems clear.

And then there was a feud between Takamatsu and Matsui. According to their later colleague Nakazawa Iwata, they did not agree on the future direction of the department: "[Takamatsu] confronted Matsui, who was weakly application-oriented, and took over responsibility for all the subjects within the Department of Applied Chemistry.... Therefore, Takamatsu was the true founder of applied chemistry in Japan."[4] Matsui resigned from the department in April 1887.[5] One month later, Nakazawa returned from overseas study in Germany and joined the department as full professor.

Nakazawa did not make clear exactly over which issue they fell out, but it is not difficult to guess. Matsui's noncommittal attitude to industrial on-site operations, shown during his overseas study at Columbia, set him apart from Atkinson and especially from Takamatsu (see chapter 3). Indeed, at Tokyo University Matsui was active more as a chemistry lecturer and analysis trainer and was largely silent on how to teach the subject of applied chemistry and chemical technology. In contrast, Takamatsu was thoroughly focused on this very question, promoting the engineering and "on-site" approach to applied chemistry teaching (see chapter 4). This chapter's opening quotation perfectly captures Takamatsu's teaching style as hands-on instruction to would-be manufacturers and consumers of chemicals, which was exactly what contemporary Japanese dyers, for example, would have needed.[6] As the Imperial University strengthened the network between the College of Engineering and the Ministry of Agriculture and Commerce and private companies by means of external student loans tied to employment obligation after graduation, it is hardly surprising that Takamatsu prevailed.

The Department of Applied Chemistry therefore experienced a gradual process of institutionalization between 1886 and 1888. It was Takamatsu who was mainly responsible for the design of the new department's curriculum; Nakazawa supported him in its execution from 1887 until 1897.[7] To understand this process, therefore, it is imperative to look at who Nakazawa was and how his pedagogical and research styles were formed.[8]

Nakazawa Iwata, German Studies, and Tokyo University

Nakazawa, like Sakurai and Takamatsu, studied chemistry at the Tokyo Kaisei Gakkō and Tokyo University with the English chemist Atkinson. But his eventual introduction to chemistry originated in his German studies that date back to 1870. It was when he received an order from his native domain of Fukui to learn German as a *kōshinsei* according to the pro-German attitude of its former *daimyō* Matsudaira Yoshinaga. This was based on the latter's high esteem for the Prussian state's military power and its high standard of general education.[9] Nakazawa was originally admitted to the German division of the Nankō in 1871, and when that was abolished in 1875, he transferred to the English division of the Tokyo Kaisei Gakkō. He entered Atkinson's Department of Chemistry there the following year and graduated in 1879 with a thesis on indigenous techniques of copper smelting. Soon thereafter he was appointed junior assistant (*jun jokyō*) in chemistry under Atkinson and assisted in Atkinson's research on sake brewing.

A turning point for Nakazawa came in 1881 when Atkinson resigned and he began to work with a new superior, the German industrial chemist Gottfried Wagener, who was teaching chemical technology and doing research on Japanese pottery. The Ministry of Education had appointed Wagener because he was familiar with the practices of the manufacturing industry in Japan (see chapter 4). Nakazawa's encounter with Wagener was therefore important not only in terms of his subsequent German connections, but also in terms of his approach to the teaching of chemical technology. Wagener's educational and career background as an industrial chemist is extraordinary by any standard.[10]

Gottfried Wagener and the Engineering Approach

Born in 1831 in Hanover, Wagener studied mathematics and natural sciences at the University of Göttingen under, among others, mathematician Carl Friedrich Gauss. Wagener was granted a Ph.D. in mathematics and physics in 1852 with a mathematical thesis related to geometrical drawing. He moved to Paris where he worked as a freelance German-language teacher and attended lectures on a variety of scientific subjects, including the chemistry lectures of Jean-Baptiste Dumas at the Collège de France.

After a short-term lectureship in the theory and practice of watch- and machine-making at a technical school in Switzerland, Wagener set up a partnership with his brother in running a chemical works. Though commercially unsuccessful, this experience took him to Japan in 1868 on the invitation of an American firm to establish a soap factory in Nagasaki. This invitation then led to his job as technical advisor at a pottery in the Saga domain, a region famous for the manufacture of luxurious porcelain, *Arita-yaki*. This is the same topic Matsui chose for his doctoral work (chapter 3), but Wagener was much more closely associated with the actual manufacturing site and process. He then embarked on a new career as a teacher and industrial chemist specializing in applying Western science to traditional Japanese pottery. Prior to his appointment as professor of chemical technology at Tokyo University, he taught the same subject at the Kyoto Seimi-kyoku, the municipal school of chemical technology and its experimental station.[11] By this time, Wagener had accumulated considerable experience in both teaching and the practices used in indigenous Japanese industries.

The approach Wagener adopted in his lectures on chemical technology was to "explain manufacturing-related theories and laws broadly and in depth according to the latest scholarly investigations."[12] This looks similar to the approach of his predecessor, Atkinson, but the following sentences show that Wagener had a different object in mind:

> The reason for adopting this approach is to enable students to understand the structure and the use of machines operated, to notice and devise ways of removing difficulties encountered in manufacturing processes, and to supervise manufacturing enterprises. Concerning recent inventions and improvements, I certainly taught only those which were already tested and had proven their usefulness when applied in practice.[13]

Wagener essentially showed here an engineering approach to chemical technology and placed more emphasis on day-to-day machine operations in factories.

Like Atkinson, Wagener assigned topics on the examination of Japanese raw materials and indigenous Japanese manufactures to his students as graduate work, but the choice of thesis topics was made with the object of using Japanese raw materials such as oil, soil, camphor and lacquer, and ores to manufacture chemistry-related products such as glass, soap, pigments for porcelain, and Japanese lacquerware more cheaply.[14] The research problems Wagener gave to students were concrete, derived directly from actual manufacturing processes, and they showed more sensitivity toward the market than Atkinson's had. Wagener also undertook experimental work on pottery himself and developed a new type of porcelain with his assistants and a potter, who had earlier collaborated with him when he held a post under the Ministry of Home Affairs.

To sum up: Wagener had not received systematic academic training in chemistry, but his doctoral studies in physics and mathematics probably helped shape his engineering approach to chemical technology. Furthermore, he had considerable on-site experience in chemistry-related factories and workshops, especially in Japanese potteries. Both his studies and his practical experience were reflected in his teaching of chemical technology in Tokyo. Nakazawa later recalled how Wagener had developed his idea of combining on-site (at potteries, in his case) and laboratory training in chemical technology at Tokyo University.[15] Wagener undoubtedly was a major influence on the development of Nakazawa's teaching style for chemical technology.

Sites of Nakazawa's Overseas Study in Germany

Nakazawa consulted Wagener concerning his choice of where and what to study overseas between September 1883 and March 1887.[16] As a result, Nakazawa studied mainly in Germany and went beyond the ordinary academic sphere of universities and *technische Hochschulen*. After attending lectures on mineral chemistry, chemical technology, and metallurgy and doing laboratory work at the University and the Technische Hochschule in Berlin for the first half year, Nakazawa spent most of his time overseas visiting industrial exhibitions, local technology-related *Fachschulen*, that is, German vocational schools, and a wide variety of factories for products, such as bricks, plate glass, textiles, pottery, beers, glass, paper (in an electrically powered plant), cement, and sugar as well as a workshop for dyeing textiles. In some of these firms, he became a *Praktikant* (apprentice).[17] The two schools he visited during this tour were the Royal Weaving, Dyeing, and Finishing School in Krefeld in the Rhineland and the Brewing School in Weihenstephan, part of the Royal Bavarian Agricultural Academy. They provided training directly related to actual manufacturing operations for future craftsmen in the textile and brewing industries, respectively, especially for designers and brewers.[18]

Nakazawa's study in Germany was thus different from that of most academic chemists including his later colleague, Takamatsu, whose study overseas was mainly based in educational institutions with only an occasional visit to chemical works. Nakazawa spent most of his period in chemistry-related factories and local vocational schools. This would be reflected in his teaching at the Department of Applied Chemistry which favored on-site factory training, and indeed in its entire curriculum.

Curriculum Development

The curriculum of the Department of Applied Chemistry was fairly consistent during the period from 1886 to 1897, and the curriculum of the academic year of 1887–1888 gives an idea of its important characteristics (table 7.1). First, laboratory work in this department was more oriented toward analytical training than was the case in the Department of Chemistry (see chapter 6). The curriculum comprised qualitative analysis, quantitative and technical analysis, determination of minerals, assaying, and blowpipe analysis.[19] Chemical analysis was of course also a basic component of the teaching in the Department of Chemistry. However, Sakurai and Divers also allocated

Table 7.1 Curriculum of the Department of Applied Chemistry, College of Engineering, Imperial University, 1887–1888

	First year	Second	Third
Mineralogy	Milne (1,2)		
Applied Physics	Shida (1,2)		
Physical Laboratory	Yamagawa (1,2)		
Steam Engine	West/Miyahara (1,2,3)		
Mechanism	Iguchi (1,2,3)		
Water Motors, Pumps, Crane, etc.	Iguchi (3)		
Building Constructions		Nakamura (1,2)	
Mechanical Drawing	Iguchi (3)	Iguchi (3)	
Organic Chemistry	Kawakita (1,2,3)		
Applied Chemistry		Takamatsu/ Nakazawa (1,2,3)	Takamatsu/ Nakazawa (1,2,3)
Applied Chemistry Lab			Takamatsu/ Nakazawa (1)
Qualitative Analysis	Shizuki (1,2)		
Quantitative Analysis	Kawakita (3)	Kawakita (1,2)	
Technical Analysis		Nakazawa (3)	
Metallurgy		Milne (1,2,3)	
Determination of Minerals		Milne/Matoba (1,2)	
Assaying			Kawano (1,2)
Blowpipe Analysis		Kawano (1,2)	
Factory Planning, Designs and Drawing			Nakazawa (1,2)
Applied Chemistry Thesis			o (3)

Note: Numbers in parentheses shows the terms (3 terms per year) in which each course opened.
Sources: Jpn. & Eng. Calendars Imp. Univ.; TDn, vol. 6.

time for organic synthesis and physico-chemical measurements. In contrast, the Department of Applied Chemistry used the whole laboratory teaching allocation of the first two years to train students thoroughly in analysis as much as in Atkinson's time, showing a continued emphasis on chemical analysis for industrial chemists in Japan.

Second, a large part of the curriculum of the first two years was devoted to engineering subjects, such as applied physics, steam engine, mechanism, water motors, pumps, cranes, etc., and it included building construction, mechanical drawing, and practice sessions for applied chemistry students.[20] Following the perceived Mancunian model, Takamatsu had already introduced engineering subjects into his curriculum for the Department of Applied Chemistry at Tokyo University in 1883 (see chapter 4). However, a large number of engineering topics was added during the first two years after the establishment of the Imperial University in 1886. These topics became the main feature distinguishing the curriculum of the Department of Applied Chemistry from that of the Department of Chemistry, where all engineering subjects disappeared after 1886 (see chapter 6).

The two full professors, Takamatsu and Nakazawa, and the two assistant professors, Kawakita and Shizuki, played different roles in this curriculum. Takamatsu and Nakazawa lectured on applied chemistry for advanced students, whereas Kawakita lectured on organic chemistry for junior students. The "purely scientific" subject of chemistry was clearly treated as preliminary and subordinate to more advanced chemico-technological subjects as had been the case in Atkinson's time. This division also permeated laboratory training at the Department of Applied Chemistry, where Takamatsu and Nakazawa supervised graduate theses, and Kawakita and Shizuki trained junior students in quantitative and qualitative analysis. The two levels were bridged by Nakazawa's course on technical analysis in the third term of the second year.

The Department of Applied Chemistry therefore had the same two-chair structure as the Department of Chemistry and also had the same division of roles between juniors and seniors. However, the nature of division was different because at the Department of Chemistry junior professors supervised advanced students' research work as well as analytical training. How did this difference come about? I return to this question later in the section on graduate work.

Strangely enough, there were no lectures on inorganic chemistry until the academic year 1896–1897, when Kawakita was appointed full professor in the department. One possible explanation is that the subject matter of inorganic chemistry was explained under the heading qualitative analysis, but it remains true that the curriculum of the Department of Applied Chemistry deviated from the standard format of academic chemistry in Europe, the United States, and Japan prevailing at that time. It is another sign of the subordinate role of purely scientific chemistry in this department in the period we are looking at.

Takamatsu and Nakazawa in the Classroom

Extant recollections by Takamatsu's former students provide some evidence about the effectiveness of Takamatsu as a lecturer. According to Inoue Jinkichi, Takamatsu's former student and colleague in the department: "He spoke clearly in lecturing; the content of his lectures was thoroughly organized, and it was easy to take notes of them.... Students were all impressed. During the three years of my study, Professor Takamatsu lectured on the subject in his charge carefully with the same attitude and tone, which satisfied students greatly."[21] Mizuta Masakichi, a later student of Takamatsu who followed an industrial career, asserted that "Professor Takamatsu's lectures were crystal clear. He explained difficult theories and complicated technical problems so simply that we listened to his lectures with great interest and could continue our studies with pleasure."[22] Nakazawa's lecturing, according to one student, "was very detailed and carefully executed. Students were a little flustered at first because of his fast speech, but we soon got used to it and had no trouble understanding it."[23] As the class size in the Department of Applied Chemistry was consistently larger than that in the Department of Chemistry, the communication skill of lecturers in classrooms were crucial (chapter 5, table 5.1 and table 7.2). Regarding lecture languages,

Table 7.2 Student enrollment in the Department of Applied Chemistry, College of Engineering, Imperial University, 1886–1900

	PG	3rd year	2nd year	1st year	Occasional	Total	Undergraduate total
1886	0	2	11	5	1	19	19
1887	0	11	5	0	1	17	17
1888	0	4	1	3	2	10	10
1889	0	1	3	2	4	10	10
1890	0	3	1	3	3	10	10
1891	0	3	2	8	4	17	17
1892	0	2	8	8	2	20	20
1893	0	8	8	9	1	26	26
1894	2	8	9	7	1	27	25
1895	1	9	5	9	0	24	23
1896	0	5	7	9	0	21	21
1897	1	7	9	13	0	30	29
1898	0	9	9	11	0	29	29
1899	4	9	10	5	2	30	26
1900	5	8	7	10	3	33	28
Average						22	21

Sources: Jpn. Calendars (Tokyo) Imp. Univ.

circumstantial evidence, such as that gleaned from Takamatsu's textbook, suggests that Takamatsu's lectures were delivered in Japanese. Nakazawa did not publish any textbook, but he did publish several educational articles in Japanese on pottery, glassmaking, and the manufacture of sulfuric acid and alkali in the *Tokyo kagaku kaishi*, which all constituted part of Nakazawa's lectures shown above.[24] In short, the department could not afford to be bilingual.

At the same time, students commonly indicated that the secret of Takamatsu's and Nakazawa's success as lecturers lay in *what* as well as *how* they communicated. Detailed treatment of subject matters as well as clarity was naturally valued by would-be technologists. Annual college reports between 1886 and 1889 describe the lectures more in detail. The professors organized their subject matter according to raw materials and final products rather than names of chemical compounds, except in the case of the Western heavy and fine chemical industries. Takamatsu thus lectured on "the manufacture of animal and vegetable oils, mineral oils, soap, dry distillation of woods, wood vinegar, wood alcohol, coal gas, charcoal gas, and oil gas" in the academic year 1887–1888.[25] Nakazawa lectured on "the manufacture of starch, artificial rubber, sugars, theory of brewing, manufactures of beer, sake and wine, lime, lacquer ash, glass etc" in the same academic year 1887–1888 and on "the manufactures of porcelain, enamel, cloisonné (*shippō*), sulfur, sulfuric acid and other acids, common salt, sodium sulfate, sodium carbonate, caustic soda, potassium carbonate, iodine, and bromine" in the academic year 1888–1889.[26] Thus, their lectures were better represented by the older subject name of chemical technology ("manufacturing chemistry") than by applied chemistry.

A comparison of Sakurai's lectures on alizarin (see the appendix) and Takamatsu's lectures on dyeing for the academic year 1888–1889 reveals the characters of both lectures. Though Sakurai mentioned the economic impact of synthetic alizarin, his treatment of the subject matter was mainly concerned with the chemical and physical properties of alizarin and the discussion of its constitution by comparing the chemical formulas of alizarin, anthracene, and anthraquinone. Most important, his lecture was silent about its manufacture and use in actual dyeing processes. In contrast, Takamatsu's lectures on applied chemistry dealt with "the distillation of coal tar, the refinement of distilled oil, manufacture of aromatic compounds of industrial importance and dyes, the structure, softening and bleaching of fibers such as silk, wool, cotton and hemp, manufactures of mordants, the properties of animal, vegetable and mineral dyes, manufacture of pigments, dyeing by soaking and printing, etc."[27] That is, they were concerned with dyeing and dye manufacture, as was his contemporary textbook in Japanese on dyeing, which was published in 1895. This was a "general discussion of dyeing, fibers, scouring, bleaching, water used in dyeing, mordants, pigment, soaking, printing, and the principles of harmony and contrast of colors coloring," suggesting that this textbook originated in Takamatsu's lecture notes used at the Imperial University.[28] The recipe-like instructions for the use of alizarin in Takamatsu's textbook convey his thoroughly practical approach to the teaching of applied chemistry, which was exactly what contemporary Japanese dyers would have needed.

Two-Tier System of Students' Graduate Work

Another aspect of the Department of Applied Chemistry that contrasts with the Department of Chemistry is the third-year graduate work. Takamatsu and Nakazawa were very directive, assigning research topics to students and supervising their work; there is no doubt that the professors were in charge. This is because students' graduate work is an extension of a full professor's extramural consultancy work, which I discuss later. Only a professor of Takamatsu or Nakazawa's experience and skill in interacting with manufacturers and industrialists could give such research topics to students.

The final project was structured in two sections: one involving a thesis and the other involving factory planning, design, and drawing. The nature and contents of students' final work therefore reveal a fusion of laboratory and on-site training. They also demonstrate a link between students' training and the development of a social network with local manufacturers in the teaching program of applied chemistry.

The research topics Takamatsu and Nakazawa assigned students for their graduate theses clearly are classified as: (1) the improvement of Japanese indigenous manufactures, such as the investigation of enamel and cloisonné; (2) the exploitation of unused natural resources in Japan, such as the method of extracting iodine from seaweed produced in Japan; and (3) solving technical problems in modern chemical industries introduced by transfer from Western countries, such as glassmaking.[29] All theses were grounded in analytical chemistry plus fieldwork that explored contemporary industry from which samples were acquired. Fieldwork aimed at "improving" Japanese indigenous manufactures by applying chemical knowledge and techniques was a legacy from the era of Atkinson, with whom both Takematsu and Nakazawa had studied (see chapter 2). Somewhat paradoxically, their choice of topics showed greater familiarity with transferred Western chemical industries, such as the alkali and synthetic dye industries, than Atkinson's because of their study of chemical technology in Europe and the experience they accumulated as consultants, as I discuss in the next section.

Sometimes students' graduate theses were published in Japanese periodicals, such as *Tokyo kagaku kaishi* (the organ of the Tokyo Chemical Society), *Kōgaku kaishi* (the organ of the Engineering Society in Japan), and *Kōgyō kagaku zasshi* (the organ of the Japanese Society of the Chemical Industry, from 1898) under the sole authorship of the student concerned. The titles of their papers, as suggested by the titles of those published in the *Tokyo kagaku kaishi* between 1886 and 1900 corroborate the above classification of research topics assigned by Takamatsu and Nakazawa.[30] Of the 35 articles written by professors or students of the Department of Applied Chemistry,

18 were concerned with the improvement of Japanese indigenous manufactures, 5 were on the exploitation of unused natural resources in Japan, and 12 intended to solve technical problems in modern chemical industries introduced by transfer from Western countries. Also noteworthy is the language they used. As can be seen from an investigation of the Royal Society's *Catalogue of Scientific Papers*, not only the students, but also the professors of the Department of Applied Chemistry rarely published in Western languages, at least until 1900. In marked contrast to the College of Science, the College of Engineering did not publish its own Western language journal until 1904.[31] This is an indication of the intended audience of the researches produced in these two departments. At least as an aspiration, the College of Science looked toward the international scientific community, whereas the College of Engineering served domestic clients.

Despite their common goal, Takamatsu's and Nakazawa's styles of research supervision were different. According to students' recollections, Takamatsu frequently visited the teaching laboratory, monitored students' work carefully, never scolded them, and ran the whole department as though he were a "patriarch (*kachō*)."[32] His "patriarchal" disposition arguably derived from his early education as a future *nanushi* (see chapter 3). Nakazawa's supervision, as recorded by students, was equally enthusiastic, but could hardly be regarded as "liberal." He was said to be rather strict, stubborn, and outspoken, and he often scolded students, hence his nickname "Dr. Thunder (*Kaminari hakushi*)."[33] This behavior was rarely seen in his English-trained colleagues, such as Takamatsu and Sakurai.

The course of factory planning, design, and drawing for third-year students most strongly reflected Nakazawa's teaching philosophy of stressing "business senses." It was initially simply called "design [*ishō*]" and was combined with the preparation of graduate theses in 1886.[34] It gained an independent existence in 1887 and was called "factory planning [*Kōjō Mokuromi*]" by Nakazawa, who took charge of this course with occasional support from Takamatsu. From this point on, it became an independent course on the design and planning of factories as a whole. According to Nakazawa,

> the object of the course of factory planning is to assign to students the drawing plans of a whole factory or part of a factory and to require them to add their own design in order to develop their business sense [*jitsugyō no nenryo*].[35]

A graduate thesis and a factory planning exercise should be well-coordinated for him. According to his 1888–1889 annual report, "I chose the manufacture of cement and its factory planning and research on the composition of glass and the planning of a glass factory as the topics of graduate theses for third-year students."[36] Chemical research on a particular manufactured product was only the start of the student's work, the end point of which was the planning of a factory built to reflect the students' R&D.

These pedagogical projects could be costly, and they were therefore financially supported by the Imperial University. The Regulation for College Research Trips (*Bunka Daigaku Gakujutsu Kenkyū Ryokō Kitei*) introduced in 1886 was one of the few sources of research money for professors, undergraduates, and postgraduates of the Imperial University. Its very nature favored fieldwork, including the collection of specimens in zoology, botany, and geology, astronomical and geomagnetic observations, and samples for the analysis of natural products. For example, Sakurai's assistant professor Yoshida Hikorokurō as well as Divers and Haga Tamemasa benefitted from this scheme for their researches in natural product chemistry.[37] The Department of Applied Chemistry took double advantage of this system. Students' and professors' inspection tours to local factories and manufacturers were needed not only for graduate theses but also for on-site teaching of manufacturing chemistry and factory design and drawing, and they were funded under this regulation.[38]

Placing Graduates, Providing Consultancy, Forging Networks

This training practice undoubtedly came from Takamatsu's and Nakazawa's intention to train their students as "chemical technologists" (*kagaku kōgyōka*), and necessitated their forming networks with chemical works and local manufacturers by means of graduates and the consulting works of professors. Annual reports of the College of Engineering for the years from 1888 to 1890, which contain three sets of data about the first workplaces of the graduates of the Department of Applied Chemistry (table 7.3), point to certain career patterns. First, the Department of Applied Chemistry produced few teachers with the exception of teachers for the Tokyo Shokkō Gakkō (Tokyo Technical School, see chapter 4).[39] Established in 1881, this educational institution was the exception in offering teaching positions to graduates of the Department of Applied Chemistry because both provided highly practical training.

Indeed, there was considerable mobility between the professoriate of the Department of Applied Chemistry at the Imperial University and the teaching staff of the Tokyo Shokkō Gakkō. For example, Takamatsu held an additional position between 1887 and 1900 at Tokyo Shokkō Gakkō, where he taught dyeing. Nakazawa also taught there between 1890 and 1891, and so did other professors of the College of Engineering at Tokyo.[40] Nakazawa was eventually appointed principal of Kyoto Higher Technical School (*Kyoto Kōtō Kōgei Gakkō*) in 1902. It was at the Tokyo Shokkō Gakkō that Takamatsu undertook an important investigation on Japanese indigo commissioned by the Ministry of Agriculture and Commerce. The study involved pilot manufactures of indigo with vegetable raw materials from Tokyo as well as from Tokushima, Shizuoka, and Kagoshima prefectures that were sampled during fieldwork; a chemical analysis of their products was performed to measure their "qualities" by quantifying indigotin. Dying tests with silk and cotton threads as well as cost estimates were also carried out.[41]

Table 7.3 on the first workplaces of graduates of the Department of Applied Chemistry also suggests that a majority of graduates became engineers in private companies or governmental agencies. According to Nakazawa:

> Twelve students graduated from this department in July 1888. Two became junior teachers at the Tokyo Shokkō Gakkō, six became engineers in the government and private sectors, two became chemistry teachers, and one established an analyst's office. The large number of graduates in this year cannot be compared to last year's, and it is my utmost pleasure that most of them acquired a position in charge of actual business.[42]

The activities of Takamatsu and Nakazawa outside university walls as consulting chemists were reflected in this trend.

In addition to the ministerial commission for work on indigo, Takamatsu did consulting work with the Tokyo Gas Company where he strengthened his connection with Shibusawa Eiichi, chairman of its board of directors.[43]

Table 7.3 Employment of the graduates of the Department of Applied Chemistry, Imperial University, 1888–1890

July 1888	
Tokyo Technical School	2
Normal School	1
Private Chemistry School	1
Naval School	1
Engineering Works Bureau, Ministry of Agriculture and Commerce	2
Osaka Hygiene Bureau, Ministry of Agriculture and Commerce	1
Tokyo Gas Company	1
Hokkaido Sulfur Factory	1
Japan Chemical Manufacturing Company	1
July 1889	
Japan Steel Company	1
Onoda Cement Company	1
Mechanical Mineral Oil Company	1
Private Business	1
July 1890	
Yokohama Municipal Gas Bureau	1

Sources: TDn, vols. 5 & 6.

On that basis, Takamatsu was invited to be an executive director of the Tokyo Gas Company in 1903 and left academia. Nakazawa's activities as a consulting chemist were also impressive. Apart from his work as examiner for several industrial exhibitions, he did consulting work with the Sulfuric Acid and Soda works of the Printing Bureau of the Ministry of Finance in Ōji near Tokyo from 1887 on and became the bureau's manager in 1891. He also held the post of manager at the Osaka Chemical Refinery between 1894 and 1895.[44] As a technical advisor, Nakazawa was also instrumental in establishing a private chemical manufacturing company, the Japan Chemical Manufacturing Company, in 1889.[45] He did other consulting work for a cement company, glassworks, potteries, and breweries.[46]

Takamatsu's and Nakazawa's Department of Applied Chemistry thus formed a close network with private as well as government chemistry-related manufactures a network that stemmed from the combination of students' graduate work, professors' consultancy work, and the careers of graduates. These networks were illustrated by the well-furnished and well-organized museum (figure 7.1). It was the flagship room of the department, together with the student laboratories that symbolized the centrality of analysis (figure 7.2).[47] The museum displayed various products and looked just like the industrial exhibitions that flourished in Japan during the entire Meiji period.[48] The museum reveals the very nature of the department as the hub of information for chemistry-related manufacturing. Though a full analysis of its consequences for the Japanese chemical industry has yet to be made, this

Figure 7.1. Laboratory of Applied Chemistry, Tokyo Imperial University, ca. 1896
Source: Ogawa 1900. Courtesy of the National Diet Library, Japan

Figure 7.2. Museum of Applied Chemistry, Tokyo Imperial University, ca. 1896
Source: Ogawa 1900. Courtesy of the National Diet Library, Japan

flow of information and manpower that started in the Meiji period seems to have complemented the local initiative of innovation Morris-Suzuki emphasized in her "social network" approach.[49]

Conclusions: Followers and Dissidents of the Takamatsu-Nakazawa Regime

The teaching program created by Takamatsu and Nakazawa was well-established by the late 1890s and became a model for the teaching of applied chemistry in Japan in subsequent years; it contributed to the education and training in Japan of chemical technologists, who were different from "ordinary" chemists. This difference was symbolized by the separation of the Society of Chemical Industry of Japan (*Kōgyō Kagakukai*) from the Tokyo Chemical Society in 1898.[50]

Some of Takamatsu's and Nakazawa's immediate successors, such as Kawakita Michitada and Tanaka Yoshio, maintained this system at the Tokyo Imperial University.[51] At the time of my research in 2004, the library of the Department of Applied Chemistry at the University of Tokyo possessed graduate theses, reports of summer on-site training in factories, and occasionally factory-design plans from 1923 onward.[52] That means that the two-tier system of graduate projects comprising theses and factory planning, the hallmark of the Takamatsu teaching system, survived at Tokyo well into the

1930s at least. On the other hand, some latecomers such as Kita Gen-itsu vehemently opposed Kawakita's approach, which owed much to Takamatsu and Nakazawa, to teaching applied chemistry. Kita preferred a more pure chemistry-oriented approach. He had been assistant professor at Tokyo Imperial University under Kawakita but found his way to Kyoto Imperial University where he became a noted research organizer in a variety of fields of industrial chemistry, such as synthetic fuel, textile, and rubber.[53]

These are the two reactions to the Meiji model of applied chemistry by followers and dissidents, which I look at more in detail in the next chapter. It suffices here to say that whether they supported or opposed the approach, few Japanese applied chemists in later generations could escape being aware of their predecessors' model.

Epilogue

Departure from Meiji Japanese Chemistry

The preceding discussion in this book yielded both theoretical and empirical outcomes. Its theoretical innovation is summed up as the creative application of the concept of contact zones to the analysis of sites for scientific practice. This theoretical innovation has two aspects, the two sides of a same coin. First, the preceding discussion elucidated the thus far undervalued aspect of sites for scientific practice, typically classrooms and laboratories, examining them as sites of intricate social interactions that were taking place within the site. Second, my project transformed the concept of a contact zone from a heuristic device into a tool for spatial analysis.

At a more empirical level, this book showed the tremendous impact of Anglo-American connections on chemistry in Meiji Japan. With this assertion I do not mean that Japanese chemistry students, school administrators, and politicians were passive recipients of techno-scientific culture from Britain and North America. By focusing on the dynamics within contact zones, I hope to have shown that exactly the opposite was the case. Nor do I mean to diminish the importance of other connections, such as German ones. I did not recount the well-known story of the paramount German impact on medicine in Meiji Japan here, but even with this omission the German presence in Meiji Japanese chemistry was too important to ignore, especially in the area of laboratory design. That said, the three major conclusions in my book, drawn from the spatial analysis of contact zones, are all related to Anglo-American connections in one way or another.

First, in comparison with its British counterpart of "liberal science model," US scientific and technological education made more conscious and explicit efforts to build new types of educational institutions that catered to industrial development, and administrators of the emerging Japanese higher education system could and did readily take this up in the 1870s. Of particular importance are engineering- and agriculture-oriented land-grant colleges, mining schools, and engineering schools that started to take shape in the 1860s. These American institutions had multiple professorships in a single discipline, in contrast to their German counterparts that favored the strong leadership of ordinary professors as institute directors.

Second, chemical education at English universities and colleges in the latter half of the nineteenth century, most notably at UCL and Owens College Manchester, which had the biggest impact on Japanese chemical education, was marked by the division of labor according to which a professor or two were in charge of lectures whereas an assistant took care of daily laboratory teaching. As a result, in contrast to the "professor-centered" lecture halls, an "assistant-centered" structure of contact zones emerged in laboratories.

Japanese chemistry students in Britain experienced this structure firsthand, and some of them constructed their own pedagogical spaces in the 1880s by reinterpreting the then fashionable German chemical institute and incorporating the British division of labor into the existing framework of the American-style multiple professorships of Japanese higher education. Tokyo's Department of Chemistry most clearly and spatially exhibited this structure. The Department of Applied Chemistry also had a similar senior/junior division of labor; however, it was somehow twisted because full professors had exclusive access to potential research topics through their consultancies. They had good reasons to be involved in students' graduate projects.

Third, the interaction between British and American teachers and Japanese students led to the institutionalization of both scientific and technological approaches to chemical education at the Imperial University in Tokyo established in 1886. The former pure-scientific chemical education was based on training in analysis, synthesis, and physico-chemical measurements at the working bench. It officially underlined the "applied science" model of technological innovation invented in Britain and the United States in the late nineteenth century, according to which innovation was a result of the application of scientific discoveries.[1] In reality, however, that pedagogical regime was sustained by full professors' strong connection with the education sector so that graduates could be readily placed into teaching positions. The latter technological approach was the combination of academic chemistry training with students' participatory observation and research in local manufactures of traditional products as well as in Western-style factories. This approach emphasized the essential role of on-site experience in technological innovation in the form of scientific reexamination and improvement of indigenous manufactures, on the one hand, and the adaptation of Western manufactures to Japan, on the other.

In this epilogue I intend to argue that all three points had a profound impact on the subsequent institutionalization of chemistry and science education at Japanese universities. As the last three main chapters have made clear, however, these features did not remain unquestioned even at Tokyo Imperial University in the Meiji period. This is true especially for the second and third points about the junior/senior divisions of labor and about scientific and technological approaches toward chemical education. My argument would be, then, that these features had serious consequences through the reactions to them of later generations of Japanese chemists. In the following, by continuing the spatial analysis of contact zones, I sketch such reactions in the subsequent imperial universities at Kyoto and Tōhoku (Sendai) to underline my argument.

Kyoto Imperial University: Cohabitation of "Pure" and "Manufacturing" Chemistry

Kyoto Imperial University was planned to include four colleges of law, medicine, literature, and science and engineering, but it opened its doors in 1897 with only a College of Science and Engineering.[2] Two departments of civil and mechanical engineering were opened that year, joined by six departments of mathematics, physics, pure chemistry (*junsei kagaku*), manufacturing chemistry (*seizō kagaku*), electric engineering, and mining and metallurgy in 1898. Kyoto's Colleges of Law and Medicine were instituted in 1899, followed by the College of Literature in 1906. The first dean of the College of Science and Engineering was therefore the de facto founder of Kyoto Imperial University. This was none other than Nakazawa Iwata, professor of applied chemistry at the College of Engineering at Tokyo and a major proponent of the technological approach to chemical education (chapter 7).[3] Nakazawa worked with his erstwhile colleague and friend Kuhara Mitsuru at Tokyo University in the planning of the college and university.[4]

The most important point in the above paragraph regarding my discussion here is that the College of Science and Engineering formed a single institutional entity. Why did Nakazawa (and possibly Kuhara) choose to proceed this way? Amano Ikuo speculated that it was due to tight financial constraints and to the fact that Tokyo's College of Science had not done well in terms of numbers of students enrolled.[5] If so, Nakazawa's concern would prove well founded: Kyoto's Departments of Mathematics, Physics, and Pure Chemistry were forced to temporarily merge into the Department of Science (*Rikagakuka*) between 1904 and 1908 at least partly due to extremely low enrollment.[6] Given Nakazawa's background, however, his decision to institute a college of science and engineering could also be construed as his statement that science was first and foremost for industrial development and should go hand in hand with engineering. It is also relevant that Kuhara expressed his attitude toward manufacturing chemistry in his earlier article back in 1882, in which he argued for the centrality of pure organic chemistry and for the unity of pure and manufacturing chemistry (chapter 4).

This "cohabitation" is not only on a piece of paper. On the one hand, at Tokyo, the Departments of Chemistry and Applied Chemistry belonged to different colleges. Their departmental spaces were physically separated, which hindered the contact of students in both departments (chapter 5). On the other hand, at Kyoto, on the initiative of Nakazawa the Departments of Pure Chemistry and of Manufacturing Chemistry were accommodated in the same building in 1898 (figure E.1, above).[7] Two new buildings of similar long shape were provided for the two chemistry departments when the

College of Science and the College of Engineering were institutionally separated in 1914 and included in the renamed Departments of Chemistry and Industrial Chemistry, respectively. But the separation was far from complete physically: The two buildings stood side by side and parallel to each other, with a shared analytical laboratory in between (figure E.1, below).[8]

This geographical proximity facilitated close contact between students of pure and manufacturing chemistry at Kyoto Imperial University. The research activity of Kuhara at Kyoto is an early case in point. He was appointed the second professor of chemistry in 1898 and the second dean of the College of

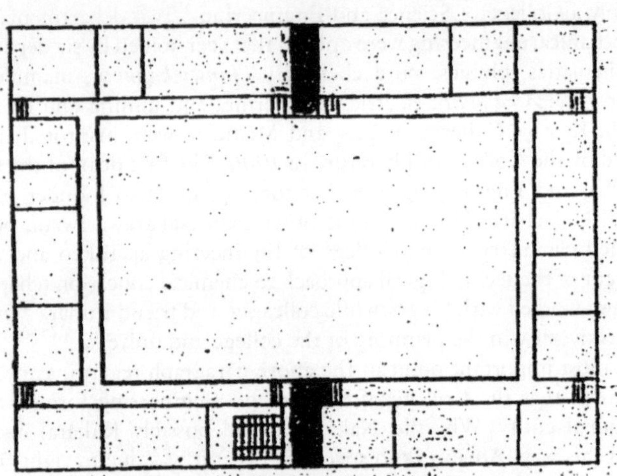

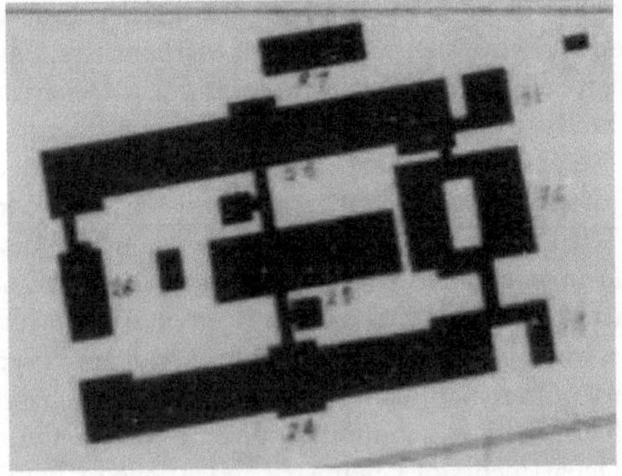

Figure E.1. Architectural drawing for the Departments of Pure and Manufacturing Chemistry, College of Science and Technology, 1898–99 (above); Buildings plan for the Department of Chemistry, College of Science and Department of Industrial Chemistry, College of Engineering, 1916–17 (below), both at Kyoto Imperial University. Courtesy of the National Diet Library, Japan.

Source: Jpn. Calendars Kyoto Imp. Univ.

Science and Engineering in 1903 at Kyoto, where he succeeded Nakazawa. There Kuhara could build a modest but fruitful research school by borrowing a managerial style, if not research topics and ideas, from Remsen: collaborating with students or assistant professors in his research and publishing results with his name as the lead author.

The most internationally renowned research of Kuhara during his tenure at Kyoto was on the reaction mechanism of the Beckmann rearrangement, an important contribution to theoretical organic chemistry.[9] However, he also took up (at least potentially) commercially relevant topics, such as the synthesis of indigo and especially arsphenamine (salvarsan), the latter of which was prompted by the blockade of German chemical products during World War I. Kuhara's research on salvarsan led to the establishment of a special chemical laboratory (*Rinji kagaku kenkyūjo*) at Kyoto in 1915 and an Institute for Chemical Research (*kagaku kenkyūjo*) in 1926.[10] Moreover, Kainoshō Tadaka, Kuhara's first collaborator in the research on Beckmann's rearrangement, later pursued a distinguished industrial career, founding the Takasago Perfumery Company (today's Takasago International) in 1920 after working with Kuhara as graduate student, lecturer, and assistant professor.[11] This mixture of scientific and industrial topics and connections was an outcome of the interaction between Kyoto's organizational style and Kuhara's deep-seated attitude toward manufacturing chemistry mentioned above.

Therefore, close connections between private industrial companies and professors of the Department of Chemistry at Kyoto seen in a later period were a natural consequence of Kyoto's style of organizing scientific and technological higher education in chemistry. For example, Horiba Shinkichi was a student of Ōsaka Yūkichi, who had been professor of physical chemistry at Kyoto's Department of (Pure) Chemistry since 1903 and succeeded him in 1924.[12] According to the recollection of Horiba's son, whenever the university was closed for holidays, he "visited his former students who were employed in various companies around the country" to make sure that they were treated properly there and also to give companies technical advice.[13]

On the other side of the same coin, Kita Gen-itsu, appointed assistant professor in 1916 and professor in 1921 at the Department of Industrial Chemistry, favored a thoroughly scientific and theoretical approach to industrial problems and to the training of industrial chemists, as noted at the end of chapter 7. His approach is best exemplified by two facts: he was appointed junior researcher (*kenkyūin-ho*) at the newly created Institute for Physical and Chemical Research (*Rikagaku Kenkyūjo* or *Riken*) upon its establishment in 1917 with recommendation from one of its founders, Sakurai Jōji;[14] in addition, he trained the first Japanese Nobel laureate in chemistry, quantum chemist Fukui Ken-ichi.[15] According to Furukawa Yasu, the geographical proximity noted above ensured exchange between the faculties of the two departments and helped Kita's curricular reform that encouraged industrial chemistry students to take subjects such as physical, organic, and inorganic chemistry offered by the nearby Department of Chemistry. A former student of industrial chemistry at Kyoto recollected that by the time of graduation, the students of the Department of Industrial Chemistry were acquainted with both teachers and students in the Department of Chemistry.[16]

Thus, Kita's approach to the training of industrial chemists was closely connected to Kyoto's institutional and spatial arrangement of pure and manufacturing chemistry created by Nakazawa. This would invite a more nuanced treatment of the attitudes of Kita and his erstwhile teacher at Tokyo, Kawakita Michitada, toward the Takamatsu-Nakazawa regime. Their attitudes were indeed more complex than it seems at first glance. On the one hand, as Divers's former student who was forced to work with Takamatsu and Nakazawa in a subordinate position, Kawakita was not keen on interacting with industrialists though he publicly advocated collaboration between academia and industry.[17] On the other hand, while staying at the Massachusetts Institute of Technology in Cambridge, Massachusetts, during his overseas study, Kita carefully studied the curriculum of chemical engineering created by application-oriented William H. Walker and incorporated it into his own teaching. Back in Kyoto, he aggressively cultivated industrial connections to fund his research school and sent his graduates to companies.[18] Sakurai recognized enough of Kita's talent and the latter's affinity to his own vision of pure chemistry to recommend him, but for Kita scientific research was not to be pursued for its own sake but to address and solve industrial problems that chemists would encounter in factories. Nakazawa and Takamatsu would surely agree with this.

Thus, if one would talk about Kyoto Imperial University's departure from Meiji Japanese chemistry, it would not be on the basis of what each of the two approaches to chemical education called for, such as the importance of pure scientific research or the relevance of on-site industrial connections. It would have to be based on their schism, the kind of problem that also frustrated Ikeda Kikunae at Tokyo Imperial University (chapter 6). Even so, this schism did not easily disappear. For example, the Tokyo Chemical Society was renamed the Chemical Society of Japan in 1921 but was not reunited with the Society of Chemical Industry of Japan until 1948 when it became a truly national chemical society.[19] The establishment of the next two imperial universities, Kyushu and Tohoku Imperial University, is another good example of the continued schism, as they opened only a college of engineering and college of science, respectively. Let's move on to the examination of one of such examples, Tōhoku Imperial University.

Tōhoku Imperial University: "Research/ Experiment-Centered" Pedagogical Space

Tōhoku Imperial University was established in 1907 by adding to the long-standing Sapporo Agricultural College a brand-new college of science, which actually did not open until 1911.[20] The faculty of the latter consisted of graduates of Tokyo's College of Science, including the three professors of chemistry, Majima Toshiyuki, Katayama Masao, and Ogawa Masakata. I have already pointed out that they adopted the *Zasshi-kai* (Journal Meeting) from Tokyo's Department of Chemistry.[21]

However, Tōhoku's Department of Chemistry was far from a copy of Tokyo's. Kaji Masanori argued that the formation of a research school in organic chemistry originated not in Tokyo's, but in Tōhoku's Department of Chemistry with Majima.[22] Firm commitment to building research schools would characterize Katayama's and Ogawa's careers as well. Katayama trained a sizable number of physical chemists in the variety of subfields, and some of them later achieved international fame.[23] Yoshihara Kenji made an interesting point about Ogawa's later maturity as research supervisor at Sendai. In contrast to his early years there (chapter 6), between 1920 and 1930 he assigned more approachable topics in the chemistry of platinum group elements and trained several doctoral candidates while Ogawa continued his research on Nipponium.[24] What counts here is not their maturity or success, but their commitment and aspiration to build research schools, something they held on to throughout their careers. My point would be that the genesis of this research-centered institutional culture at Tōhoku is understandable only by taking into consideration the dynamics at work in the pedagogical space at Tokyo.

As I discussed in chapter 5, in the 1900s the research imperative at the Department of Chemistry at Tokyo developed in the bottom-up way from junior professors, crucially Majima, rather than in the top-down way from senior professors, Sakurai and Divers. The research imperative was a by-product of the division of labor between senior professors, who were primarily lecturers, and junior professors, who were in charge of experimental training. This division had an impact on the space of the chemistry department at Tōhoku, where Majima was appointed the first professor in organic chemistry in 1911 (figure E.2).

Looking at the blueprint of Tōhoku's Department of Chemistry, one will notice that there were separate buildings for experimental training and for lectures. All offices of full and assistant professors were close to student laboratories, regardless of their rank.[25] In a sense, this was a copy of the student laboratory area at Tokyo, with which chemistry professors at Tōhoku were all familiar. In contrast, unlike in Tokyo, lecture rooms in Tōhoku were separated from laboratory areas and professors' offices. My argument here is that this experiment-centered structure of Tōhoku and its research-centered institutional culture reflected the sometimes strained experiences of Tōhoku's chemistry professors while they were students and junior professors at Tokyo. This generational divide was in the main about place and role divisions in Tokyo's pedagogical regime, which was carried over neither to Tōhoku's nor to Tokyo's new chemical institute, both built in the 1910s (chapter 5).

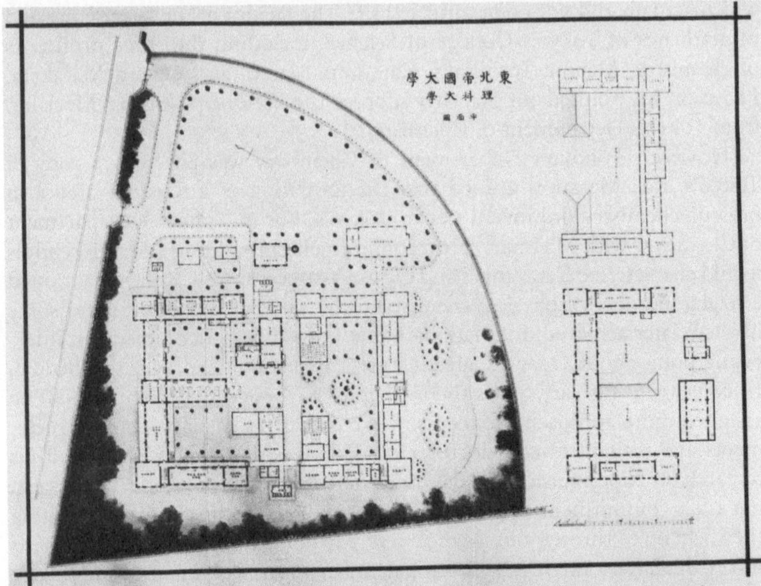

Figure E.2. Architectural drawing for the College of Science, Tohoku Imperial University, 1913–14. Note that the left building was for experimental training, highlighted by benches, and the right side for lectures designated by narrow desks in lecture halls. Courtesy of the National Diet Library, Japan.

Source: Jpn. Calendar Tohoku Imp. Univ. Coll. Sci.

The *kōza* (chair) System and the Pedagogical Space of Chemistry

What Tōhoku's and Tokyo's new chemical laboratories tell us is what might be called a "flattening" effect of intergenerational conflict between the first (Meiji) generation and second (Taisho) generation of Japanese chemists on pedagogical spaces. This situation led to the transformation of the second "division of labor" aspect of Meiji Japanese chemistry teaching into a further decentralized, fragmented departmental structure in the 1910s, where many full professors taught and did research *on equal footing* and largely independent from each other, each in his own "fiefdom." I emphasized "on equal footing" as this has often been taken for granted. But chapter 5 of this book shows that this was not always the case, especially at Tokyo Imperial University in the late Meiji period. I would argue that this is part of the larger development of the Japanese higher education during the interwar period into what has been misleadingly called the "*kōza* (chair) system," with each *kōza* made of one full professor, one assistant professor, and one assistant, which has ever since defined the structure of the Japanese scientific pedagogical and research system even well after World War II.[26]

With my arguments in this book, I contend that the seemingly "national" institution of the *kōza* system was actually an often unpredictable outcome of development in scientific pedagogy that had been inspired by, but was different from, American, British, and German precedents rather than the direct outcome of a decree stipulating the establishment of a *kōza* around 1890, as conceived and instituted by bureaucrat Inoue Kowashi. Decrees solidify and sometimes have impact on pedagogical realities but rarely create them. Terasaki Masao and Amano Ikuo have already made a couple of important points that support my approach.[27] First, the numbers of chairs in each college and department were determined by (sometimes tough) negotiations between the Imperial University and Inoue. As Terasaki convincingly showed, this means, on the one hand, that the outcome reflected Inoue's stance favoring pure over applied subjects, but on the other hand, it also means that Inoue could not ignore the reality of university teaching. In both pure and applied chemistry at Tokyo, for example, Inoue's decision simply corroborated the existing two-chair system. Second, the original *kōza* was very different from "the *kōza* system": A single professor would officially be in charge of each chair. The subordinate and supportive roles of assistant professors and assistants were stipulated separately in the same Imperial University Ordinance, but their numbers were much lower than those of full professors.[28] As Amano pointed out, beyond that was largely the realm of usage (*un-yō*), and in many cases an assistant professor occupied a chair. Many aspects of actual usage have not been clarified yet.[29]

There are at least three major events in the development of chemistry teaching detailed in this book that affected the genesis of the *kōza* system in addition to what I have already pointed out, i.e., the transformation of the second "division of labor" aspect of Meiji Japanese chemistry into a decentralized departmental structure in the 1910s. The first is the adoption at the Tokyo Kaisei Gakkō of several professorships for each department following the American model. This dates back to 1873 when Griffis made a proposal to Tokyo Kaisei Gakkō and Ministry of Education officials to that effect (chapter 2). This adoption was solidified by the second event, namely, the redesign of the "general" chemical laboratory, based on the German one-chair system, into the main building of the College of Science at (Tokyo) Imperial University based on the two-chair system between 1885 and 1888. Third, with Divers and Kuhara as important pioneers, there was an increased level of group research—problem assignment and collaboration—between full professors, assistant professors, lecturers, and graduate students in the 1910s, as is shown by the example of the three Tohoku chemistry professors.

The next obvious question is how far the importance of such events extends beyond my case studies. We have reasons to be cautious. One major issue I could not address is the relationship between the *kōza* system in science and its powerful relative, medicine. Terasaki's warning that any historical discussion of the *kōza* system should pay due attention to disciplinary differences is still a valid one, especially if, as suggested by Amano, many aspects of the *kōza* system rely on its usage in a variety of disciplinary and subdisciplinary settings.[30] Nevertheless, chemistry's fairly early start at institutionalization in Japanese colleges and universities and its pioneering role in the construction of disciplinary and departmental spaces suggests that it played a major part in providing a reference point for the development of teaching and research regimes in other disciplines.[31]

How does the above discussion affect the evaluation of this "Japanese" system? The background of this question is that it was not an ideal solution in everyone's eyes and drew criticism from a wide variety of audiences. Among these were scientific advisory groups in the United States and "democratic" reformers of academic hierarchy such as the physicist Sakata Shōichi at Nagoya Imperial University right after the end of World War II, dissident Japanese academics in the "age of activism" in the 1960s, and historians.[32] These critics have rightly pointed out that the *kōza* system lacks "scrap and build" flexibility, that the freedom of research for entry-level researchers is often discouraged, and that it impedes the emergence of a leader within a department and discourages the coordination of research efforts toward a large project. Yet there has been another set of voices, less vociferous than those of critics or dissidents but equally revealing, in the discussion of the *kōza* system in Japan; according to these other voices the system was workable, relatively stable, and had its own strengths.[33] Arguably, it ensured the balanced long-term development of a discipline as a whole amidst changing political, economic, and intellectual trends. The development of, say, physical chemistry in Japan was sustained by the existence of a *kōza* earmarked for it and did not overwhelm other subdisciplines of chemistry, especially outside Tokyo.

Based on my case studies, I would argue that the claim that the *kōza* system was inherently "feudal" is misplaced; rather, if I am correct, it partly came from the "flattening" effect of intergenerational conflicts between Meiji and Taisho chemists. Any historical appraisal of an institution should be based on the context in which it was situated. More important, the system has facilitated close "disciplining" and the effective transmission of problems, skills, and values from senior to junior researchers within a "fiefdom." This is obvious in the close connections between the genesis of the *kōza* system and increased group research that I have pointed out above. Thus, arguably, the *kōza* system encouraged the stable production of quality, if not always highly original, scientists and engineers, a prerequisite for massive technological development of Japan throughout the twentieth century. This story of Japanese chemistry tells us an important part of the system's origins.

Appendix: An Excerpt from Sakurai's Lectures on Organic Chemistry and the Corresponding Part of Majima's Notebook of Sakurai's Lectures

Sakurai's Original Lecture Notes

Alizarine $\left[C_6H_4 : \begin{matrix} CO \\ CO \end{matrix} : C_6H_2(OH)_2\right]$ $C_{14}\ H_8O_4$.

Alizarine is a well known red colouring matter extracted from the roots of 'madder' plant, botanically known as Rubia tinctoricum. The colouring matter, however, does not exist as such in the plant, but in the form of a glucoside. This glucoside called *rubian* or *ruberythric acid* undergoes hydrolysis under the influence of a ferment, called *erythrozym* [sic] also cont[aine]d in the plant. No action takes place so long as the plant is living, but when it is cut and the roots are left in a moist condition, alizarine separates out.

$$\underset{\text{Ruberythric acid}}{C_{26}H_{28}O_{14}} + 2H_2O = \underset{\text{alizarine}}{C_{14}H_8O_4} + \underset{\text{glucose}}{2C_6H_{12}O_6}$$

The above equation represents the change according to Graebe and Liebermann.

Alizarine is almost insoluble in water. It dissolves in alkali forming a beautiful violet colour, from which carbonic acid precipitates it. The p[reci]p[ita]te on drying and sublimation forms fine orange-red needles. M[elting] P[oint] 290. It dissolves easily in alcohol, ether, benzene, etc. The alkaline sol[utio]n treated with

aluminium sulphate gives a red lake. (carmine red)
ferric salts give purple.
stannic salts give red (orange red)
ferrous salts give brown.

On account of the production of these comp[oun]ds alizarine has long been employed as one of the most important dyes, and for this purpose “Madder” plant was extensively planted. The extent to which madder plant was cultivated 20 years ago may be stated on an average to amount to an annual crop of 70,000 tons, valued at £ 3,150,000 a roughly 20,000,000 yen. At present the plant is scarcely grown, and yet we use more alizarine than we used 20 years ago, all this being now artificially produced, the starting point being the waste product of coal gas manufactory [sic], viz. anthracene.

The fact that alizarine is an Anthracene derivative was clearly established by Graebe and Liebermann in 1868, not very long after anthracene itself was definitely settled to have the composition $C_{14}H_{10}$. They found that by heating alizarine with zinc dust anthracene was produced.

$$C_{14}H_8O_4 + 5Zn = Zn(OH)_2 = C_{14}H_{10} + 6ZnO$$

As zinc dust always contains zinc oxide and zinc hydroxide, the latter must have been decomposed by zinc, at a high temp[erature]: $Zn(OH)_2 + Zn = 2ZnO + H_2$, thus supplying necessary hydrogen for the conversion of alizarine to anthracene.

Comparing the formulae of anthracene, anthraquinone, and alizarine they came to regard alizarine as a dihydroxy-derivative of anthraquinone:

$C_{14}H_{10}$	$C_{14}H_8O_2$	$C_{14}H_8O_4 = C_{14}H_6(OH)_2O_2$
Anthracene	Anthraquinone	Alizarine.

This view was very soon confirmed by synthetical [sic] formation of alizarin from anthraquinone.

Source: “Organic Chemistry Vol. VIII, Joji Sakurai,” Sakurai Jōji Papers C, Ishikawa Kenritsu Rekishi Hakubutukan.

Notes Taken by Majima

Alizarin ハ Rubia tinctorum あかね中ニ glucoside トシテ存在ス. 此草ヲ切リテ其根ヲ水ニ浸シテ[□□□□] ferment ノタメニ Hydrolysis シテ Alizarin ヲ生ス: $C_{26}H_{28}O_{14} + 2H_2O = C_{14}H_8O_4$ [sic] 全テ茜ヨリトリタルモノナリシ [□□]英国領印度ニテ大ニ之ヲ培養シ七万噸ニ達シ金額 20,000,000 ニ達セリ. 今日其需要ハロノ拡張ニ達セルガ今ハ皆人造ニシテ其ノ土地他ノ有要植物ノ耕作ニ供セラル.

水ニハトケヌ Alkali ヲ加ヘテトカシ, Al を加ヘル法デ赤 $FeCl_3$ ハ茶色ヲ呈ス. 即チ Alum[□□]mordant シテ alizarin ヲツケ[□□]ヨシ

1868年ニ人造法発見サル. Anthracenノ coal tar 中ニアルヲ発見ス 1866 頃ナリ. Alizarin ヲ Zinc dust ト h スレバ Antracen トナレリ.

$C_{14}H_8O_4$ カ red サレテ anthracen トナレリ. 故ニ Glaebe 等ハ其ノ Anthraquinonノ di OH deriv ナラント考ヘタリ. 而シテ直ニ confirm サレタリ

Note: □ indicates an unclear Japanese letter in the original.

Source: "Organische Chemie von J. Sakurai, Zweiter Band: Verbindungen der Aromatische Reihe," in *Majima Bunko* (4/M394), Osaka Daigaku Fuzoku Toshokan.

Notes

Introduction

1. There have been broadly two approaches in the prior scholarship in the history of science and technology in modern Japan and East Asia. One of them focuses on the translation of scientific terms into Japanese and how Japanese intellectuals made sense of Western scientific concepts (Yoshida 1974, Tsukahara 1993, and Tsukahara 1994). The other approach analyzes the building of institutions pertaining to Western science, such as universities and research laboratories, and its contribution to the formation of a research tradition in Japan (Bartholomew 1989). More or less the same pattern is seen in the historiography of modern science in China (Reardon-Anderson 1991, Wright 2000, and Elman 2008). One important exception is Fan 2004, which draws on the contact zone and cultural borderland as a "heuristic device" (Introduction, notes 18 and 22 below).
2. See, for example, Schaffer et al. 2009. Secord 2004 is a programmatic proposal for a history of global knowledge circulation in science. Also relevant are a growing number of historical studies on the impact of international travelers on the development of science in various parts of Europe, such as Simões, Carneiro, and Diogo 2003.
3. For the general discussions of science, localities, and spaces including laboratory studies, see Hannaway 1986, Hashimoto 1993, Smith and Agar 1998, Galison and Thompson 1999, Chambers and Gillespie 2000, Henke and Gieryn 2008, and Kohler 2008.
4. Gordon 2009: 51.
5. Cobbing 1998: 28.
6. See ibid.: 29–38, where Andrew Cobbing presented useful data about the breakdown of early Japanese overseas travelers and students based on country destinations that I use in a later paragraph.
7. See, for example, Lundgren and Bensaude-Vincent 2000, Warwick 2003, Kaiser 2005a, Kaiser 2005b, Jackson 2006, Taylor 2008, Seth 2010, and Heering and Wittje 2011.

8. Josep Simon made the interesting point that cross-national comparison has long been a key method for educationists and historians of education since the nineteenth century, whereas recent historians of science and technology have been more focused on the local aspect of scientific pedagogy and stayed away from this method under the influence of social constructivism (Simon 2012: 251–252). See also Simon 2011 for an excellent example of a cross-national history of scientific pedagogy with a focus on textbooks.
9. Under chemistry here I include both alchemy and early modern chemistry (chymistry) in addition to modern and contemporary chemistry. For a survey of the history of chemistry, see Brock 1993. For alchemy and chymistry, see Principe 2013.
10. Homburg 1999 and Klein 2008, as part of Kohler 2008.
11. Knight 1992 for the European image of chemistry. For "applied science" in the nineteenth and early twentieth centuries, see Bud 2012. For international perspectives on the contemporary discussion of "technical education," see Bud and Roberts 1984: 11–17 and Lundgreen 1990. For Anglo-Japanese comparisons, see Gospel 1991. For American perspectives, see Seeley 1993, Lécuyer 2010, and Angulo 2012.
12. For chemistry's impact on practical physics, see Brock 1998. For the impact of synthetic chemistry on physics, biology, and physiology, see Pickstone 2001: 135–161.
13. See note 4 above.
14. Bartholomew 1989: 65 (on hired foreigners) and 71f. (on overseas students). Cobbing cited a calculation by the Japanese Ministry of Education in 1871 that "there were as many as 107 students in Britain alone." (Cobbing 1998: 36). See also note 5 above. For hired foreigners, see also Umetani 1971, Jones 1980, and Checkland 1989.
15. See, for example, Nakayama 1978: 46–49. For subsequent German impacts on Meiji Japanese medicine, with a particular focus on the bacteriologist Kitasato Shibasaburō, see Bartholomew 1989.
16. It is difficult to substantiate this claim before the advent of science metrics in the early 1970s, and even then it remained problematic due in no small part to difficulties with definitions and the setting of national boundaries. However, Japan's chemistry became noticeable in the world in available early data. See, for example, *Science Indicators 1972* (1973): 103–105. For a reflective account of science metrics by STS scholars involved, see Elkana et al. 1978 and Thackray et al. 1985: 1–8.
17. Pratt 1992: 6. See Ortiz 1995: 97–103 for transculturation. Its sexual connotation is clear from Fernando Ortiz's own gloss: "[the] union of cultures is similar to that of the reproductive process between individuals: the offspring always has something of both parents but is always different from each of them." (ibid: 103). These two concepts are getting currency in various fields of the humanities and social

sciences, such as language education, Japanese art and museum studies, and slavery studies (Miller 1994, Morishita 2010, and Hörmann and Mackenthun 2010).

18. Pratt 1992: 7. Compare with Fan 2004: 3–4, which discusses the promise (blurring rigid cultural boundaries) and pitfall (invoking "a rosy picture of free trading of cultural production") of using "contact zones."
19. Galison 1997: 781–844. See also Gorman 2010 for the strength of trading zones in analyzing interdisciplinary collaboration in scientific and engineering research.
20. Gordon 2009: 72–73 and Herzfeld 2002. Science's role in territorial, economic, and cultural colonialism has drawn much attention, as exemplified in MacLeod 2000.
21. Mody and Kaiser 2005: 385–388.
22. For the contact zone as a heuristic device, see Fan 2004: 3.
23. Prown 1993: 1, as cited in Wittner 2008: 6.
24. See Michel Foucault's analysis of the panopticon (Foucault 1979: 200–203). See also Markus 1993: 39–168 and Smith and Agar 1998: 1–23.
25. See, for example, Bracken 2013.
26. Schaffer et al. 2009. It is less important whether my "cultural mediators" can be strictly regarded as "go-betweens" than to point out parallels here. Both resided in two worlds or more, in my case the world of teachers and that of students, and the world of British or North-American scientific cultures and that of Japan.
27. Ōhashi 1993.
28. There has been considerable discussion of how innovative and influential Liebig's Giessen laboratory was, as summarized in Jackson 2006: 283–284, and 286. One test case that has received nuanced treatment is the establishment of the RCC in London with Liebig's involvement and his most celebrated student, August Wilhelm Hofmann, as its first professor. Against the backdrop of the then prevalent "founder's myth," Gerrylynn Roberts, a social historian of British chemistry, deemphasized RCC's Giessen legacy and instead underlined localized conditions in which Hofmann had to operate (Roberts 1976). Catherine Jackson, a historian of chemistry more concerned with pedagogy and academic research in organic chemistry, also deemphasized the long-term effect of Liebig at RCC: "although Hofmann initially recreated the Giessen system in London, both he and his school at the RCC ultimately attained independence from Liebig" (Jackson 2006: 295).
29. See, for example, the contributions of Michelle Hofmann, Steven Turner, and Richard L. Kremer on the introduction of teaching laboratories in late nineteenth and early twentieth century Canadian and American secondary and higher science education, in Heering and Wittje 2011: 177–205, 207–242, and 243–280. For the social

history of chemistry in the whole of Europe during the long nineteenth century, see Knight and Kragh 1998.

1 Japanese Chemistry Students in Britain and the United States in the 1860s

1. The Tokugawa Shogunate was the de facto government of Japan between 1603 and 1868 controlled by the Tokugawa family, its trusted retainers, and close allies. *Han* were semi-autonomous domains owned and ruled by feudal lords, called *daimyō,* that were deemed subjects of the Tokugawa family. *Han* existed throughout the Tokugawa period and were abolished by the nascent Meiji government in 1871; this was a decisive step toward the creation of a centralized nation-state. It also had strong impact on higher education (chapter 2). See Gordon 2009: 61–75.
2. Major studies of this topic, mostly from Japanese history, include: Inuzuka 1974 (English translation: Cobbing 2000); Fujii 1989; Checkland 1989; Ishizuki 1992; Cobbing 1998; and Inuzuka 2001.
3. On Williamson, see Harris and Brock 1974.
4. The roles of assistants in leading teaching laboratories for chemistry in nineteenth-century England are drawing the attention of historians. See, for example, Gay 2000 and Jackson 2006: 299–301. The cases of UCL and Owens College Manchester are discussed later in this chapter and in chapter 3 (Takamatsu Toyokichi's overseas study at Manchester) of this book.
5. Introduction, note 11.
6. For an overview of the history of Dutch learning and Western learning in Japan, see Yoshida 1984. On chemistry in Dutch learning, see Tsukahara 1993.
7. On Western learning in the Chōshū domain, see Ogawa 1998.
8. Ishizuki 1992: 49–56, and Inuzuka 2001: 11–44.
9. Cobbing 2000: 76.
10. Cobbing 2000: 20–22, 81–82, 86–89, and 108f.
11. Cobbing 1998: 103.
12. Sugiyama 1981.
13. Matheson 1899: 204.
14. Foster 1905: 616. Cited in Harris and Brock 1974: 123.
15. Fujii 1989: 78.
16. UCL Council Minutes, April 13, 1861, May 4, 1861, November 1, 1862, and July 15, 1865. UCL Records Office.
17. Harte and North 1991: 83.
18. UCL Council Minutes, March 4, 1865. UCL Records Office.
19. UCL Council Minutes, November 7, 1863. UCL Records Office. According to Lahiri's calculation, the number of Indian students in Britain in 1873 was 40–50 in comparison with a larger number of

Japanese students in Britain in the same period, which amounted to over 100. Lahiri 1995: 12–15, and Cobbing 1998: 37.

20. *The East-India Register and Directory, For 1811 corrected to the 18th December, 1810* (London: Printed by Cox and Baylis, n.d.), xiv; *East-India Register and Directory, for 1819, 2nd Edition* (London: W.H. Allen, 1819), xiii; and *East-India Register and Directory, for 1834, 2nd Edition* (London: W.H. Allen, 1834), xiii.
21. Harris and Brock 1974: 97f.
22. Harris and Brock 1974: 109.
23. Williamson 1849: 12.
24. Williamson 1849: 11, 13, and 14. His anti-French feeling was made more explicit in his political tract coauthored with his father-in-law: Key and Williamson 1858.
25. Bud and Roberts 1984: 71–86 and Roberts 1998: 107–119.
26. Williamson's syllabuses and examination papers changed little between the mid-1860s and the early 1880s. See, for example, *UCL Calendar 1877–1878*: 41–44, ii–iii, xxii–xxiii, and lxvi–lxvii.
27. Williamson 1870: 9 and 11.
28. Williamson 1870: 18. Compare Williamson's evidence at the Devonshire Commission on June 24, 1870, q. 1173, reprinted in *Irish University Press Series of British Parliamentary Papers. Education: Scientific and Technical 2* (Shannon: Irish University Press, 1969), 112f. There he made another point about the impossibility of duplicating and updating chemical plants in universities.
29. Chapman 1910: 677.
30. UCL Register of Students, Sessions 1861–62, 1862–63, 1863–64, UCL Records Office.
31. University of London 1912: 475 and 501.
32. UCL, *Register of Students, Sessions 1861–62, 1862–63, 1863–64*; *Register of Students, Sessions 1866–67, 1867–68, 1868–69*. UCL Records Office. Interestingly, chemical laboratories also functioned as a contact zone between Chōshū and Satsuma students, who, after initial parochial hostilities, used them as a meeting point and socializing space. Inuzuka 2001: 117.
33. *UCL Calendar 1865–66*: 55.
34. My translation, from the letter from Masaki to Inoue, December 10, 1872. *Inoue Kaoru Monjo*, Kokuritsu Kokkai Toshokan. Cf. Numakura 1988: 38 and 46.
35. Itō's recollection cited in Inuzuka 2001: 93f.
36. Inuzuka 2001: 120.
37. Inuzuka and Ishiguro 2006: 31. For my analysis of the identities of the two teachers, see Kikuchi 2009: 123–127.
38. Inuzuka and Ishiguro 2006: 27.
39. Nishimura 1977: 9. The reading adopted in this transcription is "*Furegudekuku*" in *katakana*, but it can be read as "*Furederiku*," considering the forms of letters used.

40. See the entry of Frederick Settle Barff in Lightman 2004: 114–115. He was well known within the UCL circle for his research on metal corrosion, which Williamson took up as a topic of industrial research.
41. Kiyonari Yoshida, Certificate of a Candidate for Election, the Chemical Society, balloted December 5, 1872. The Royal Society of Chemistry.
42. Nishimura 1977: 11f. *The Times*, August 2, 1865: 9, cited in Cobbing 2000: 71–75; on 74f. Japanese visitors to Britannia Ironworks, Bedford, July 29, 1864, Bedford Museum. For the enterprise of the Britannia Ironworks in Bedford, see Knight 2001: 24–26.
43. Kikuchi 2009: 127f.
44. Italics added. Letter of A. W. Williamson to C. C. Atkinson (the secretary of UCL, not to be confused with Robert William Atkinson), July 24, 1865. College Correspondence, UCL Library Services, Special Collection. Cited partly in Checkland 1989: 282.
45. Harris and Brock 1974: 99.
46. "Plan of the Distribution of the Departments within the Building: 1828–1879" (not paginated), in Bellot 1929.
47. Tanaka and Takata 1993.
48. My translation, letter from Itō in London to Ōkuma Shigenobu (*sangi*, state councilor), Ōki Takatō (head of Ministry of Education), and Inoue Kaoru (*ōkura taiyū*, deputy head of Ministry of Finance) in Tokyo, November 4, 1872, in "Itō Fukushi narabini Rondon Hakushi Chiyaruresugurahamu Kengi" (2A-009–00, Kō 00781100: 003), *Kōbunroku*. Cited in full in Shunpo Kō Tsuishōkai 1940, vol. 1: 675–678, on 677. Ishizuki 1992: 219 cited a part through Itō's biography without clarifying the identity of *Chiyaruresu Kurahamu*.
49. "Itō Fukushi narabini Rondon Hakushi Chiyaruresugurahamu Kengi," *Kōbunroku* (note 48 above). Although Itō's letter quoted above appeared in his biography, the Graham Proposal was not transcribed there and therefore was not used by later historians including Ishizuki. In the following I cite this source from my translation of the contemporary Japanese translation of the Graham Proposal.
50. As shown by Ōki's reply to Itō in March 1873, in "Zenjō 4 ken Fukugi Jōshin" (2A-009–00, Kō00781100: 005), *Kōbunroku*.
51. Ishizuki 1992: 251.
52. For reasons for my adoption of "science" as the translation of *gakujutsu*, see note 62 below.
53. The Graham Proposal (cf. note 49 above).
54. Ibid.
55. Satō 1980: 23 et passim. Tsukahara Tōgo also argued that "scientific theory and technical practice were merged in *Rangaku*. This tradition was a remarkable feature of science in Japan" (Tsukahara 1993: 5).
56. See note 32 above.

57. "Memorandum. Japan. F.O. July 1 1864," FO46.49, The National Archives. Japanese translation cited in Inuzuka 2001: 108.
58. Letter from Itō to Inoue, March 28, 1866 (Japanese calendar), cited in Inuzuka 2001: 125.
59. Gooday 1989, chapter 4, and Gooday 1990.
60. See the photograph of Inoue Masaru with a shovel, symbolizing his apprenticeship in mines, in Murai 1915. For his life see Murai 1915 and Yamamoto 1997. Inoue's award of a certificate of honor in geology (in the name of "Nomuran") is recorded in *UCL Calendar 1867–1868*: 60.
61. See notes 33 and 34 above. In the course of analytical chemistry, Charles Graham "of Berwick-on-Tweed" was in second place.
62. *Gakujutsu* here should not be confused with the same word in modern Japanese, which means scholarly or academic studies in general. *Gaku jutsu* as a composite term made up of *gaku* (learning) and *jutsu* (arts or techniques) has a long history in Japanese language dating back to the seventh and eighth century. See, for example, *Shoku nihongi ni* (1990): 113. *Gakujutsu* continued to be used this way at least up to the early Meiji period, for example by Nishi Amane (see chapter 4, note 84). I am indebted to Ōkubo Takeharu for this point and references.
63. Ōkubo 1972: vol. 2, 45–59, on 51 and 55.
64. For the influence of Harris on the mentalities of Satsuma students, especially Mori, see Hall 1973: 95–128.
65. Hall 1973: 95–128.
66. Rutgers Scientific School, Records, vol. 1, BSC 1868–1895, Rutgers University Archives. I am indebted to Gerrylynn Roberts for this information.
67. See, for example, Veysey 1965: 70–71; Geiger 1986: 5–7; and Rossiter 1975: 139–140, 144, 160 and 228 (note 31).
68. For the projected course in chemistry and agriculture at Rutgers Scientific School, see, for example, *Seventh Annual Report of Rutgers Scientific School, for the Year 1871* (Trenton, NJ: Murphy and Bechtel, 1871), pp. 17–18. For Murray's idea of the applicability of science to agriculture, see Murray's letter to Mori Arinori dated March 7, 1872, in Ōkubo 1972, vol. 3: 357–378, on 360–361.
69. William Rieman III, "A History of the Rutgers School of Chemistry" (1972), 17. Unpublished typescript, Rutgers University Archives.
70. Demarest 1924: 376–378.
71. Sidar 1976: 105–106.
72. Ibid.: 106.
73. On Murray, see Yoshiie 1992 and Yoshiie 1998.
74. Inuzuka 1987: 229–231. Inuzuka inferred that Hatakeyama studied law, politics, literature, and religion at Rutgers based on the title list of Hatakeyama's book collection. There is no evidence supporting this inference in the Rutgers registers of students for the period

that Hatakeyama was there (note 66 above). His book collection was more likely to be used in his private study and religious life rather than in his formal study at Rutgers Scientific School.

75. Demarest 1924: 1–74.
76. Inuzuka 1987: 223.
77. "Visits to Chemical Works," *Chemical News* 15 (1867): 71. The writer of this article mentioned the winter semester in 1866/67 as the time when this practice started.
78. For his expertise in this area, see Graham 1873–1874 and Graham 1879–1880. Fruton 1985 includes Graham's name.
79. Donnelly 1997: 136–138.
80. Graham 1878–1879.

2 American and British Chemists and Lab-Based Chemical Education in Early Meiji Japan

1. Gordon 2009: 72–73.
2. See a similar comment from Tessa Morris-Suzuki on Meiji industrialization in general (Morris-Suzuki 1994: 85).
3. Nakayama 1978: 19 and 77–88; Morris-Suzuki 1994: 81.
4. The Bansho Shirabesho has been discussed in earlier works in English on the history of science in Japan such as: Koizumi 1973: 22–27 and Bartholomew 1989: 30, 38, and 40.
5. See, for example, Doak 2007: 41–45.
6. Quoted from Doak 2007: 41. The impact of Perry's arrival on the establishment of the Bansho Shirabesho has been pointed out by historians of Western learning in Japan. See, for example, Numata 1989: 185–189.
7. Hara 1992: 39–40 and 59–61. For primary sources, see Kurasawa 1983: 115–116.
8. Kurasawa 1983: 92–95, 195 and 291; *TDh, Shiryō 1*: 6–8. *Gei jutsu* here should not be confused with the same word in modern Japanese, where it refers to art such as fine art and music. Just as *gakujutsu* mentioned in chapter 1, *gei jutsu* as a term meaning "learning and arts (or techniques)" has a long history in Japanese language dating back at least to the seventh or eighth century. See, for example, *Shoku nihongi ichi* (1989): 73. The term was popularized in the Bakumatsu period by politician and Dutch scholar Sakura Shōzan with his phrase "*tōyō dōtoku seiyō gei jutsu*" (Eastern morality, Western learning and arts), see Jansen 1989: 241. It is most probably in this latter context that the writer of this draft used *gei jutsu* to refer to Western learning. I am grateful to Ōkubo Takeharu for pointing out pre-Bakumatsu usages of *gei jutsu and gakujutsu* and their references.
9. Kurasawa 1983: 292–309.
10. Koizumi 1973: 26–27; Bartholomew 1989: 32.

11. Tsuji 1882: 69–72. See also Tanaka 1964: 390–392; and Okuno 1980. For Tsuji's role in educational policy making in the early Meiji period, see Sekiguchi 2009.
12. Bartholomew 1989: 32 and Beukers et al. 1991.
13. Beukers et al.1987, Shiba 1993, 2000, and Shiba 2006: 39–64.
14. *TDh, Tsūshi 1*: 92–98. For the origin of the name Nankō and its sister school of medicine, Tōkō, see Amano 2009: vol. 1: 20–21.
15. *TDh, Tsūshi 1*: 175–176; Kurasawa 1983: 292.
16. *TTDg*, vol. 1: 152. See also Amano 2009, vol. 1: 21.
17. *TDh, Tsūshi 1*: 175–176; Kurasawa 1983: 292.
18. Griffis 1900 remains a readable biography of Verbeck still cited by historians. See also Ōhashi and Hirano 1988 and Murase 2003. Verbeck's many-sided activities as hired foreigner of the Meiji government were outlined in Umetani 1971: 30–34.
19. Griffis mistook Verbeck's alma mater as "the Polytechnic Institute of Utrecht," which never existed, and asserted that he came "especially under the care of Professor Grotte" (Griffis 1900: 47). Murase Hisayo doubted the authenticity of Griffis's statement in its entirety (Murase 2003: 63). However, in my opinion Murase has gone too far because there was indeed a German, Peter Dietrich Grothe, who was director of the Technical School in Utrecht. It was a secondary-level school conceived "by the alliance of manufacturers and university professors" to fill "the gap between elementary education and higher technical education" (Lintsen and Bakker 1994: 94). Grothe taught mathematics, physics, mechanics, and applied chemistry there between 1850 and 1864, when he became professor at the Delft Polytechnic School. It is therefore likely that Verbeck attended the school or had at least heard about it from him. See the entry of Grothe in Molhuysen 1911–1937, vol. 7: 504–505. I am indebted to Ernst Homburg for this information.
20. My translation. Kurasawa 1983: 291 and *TDh, Tsūshi 1*: 174f.
21. An early planning draft for a Kōbu Gakkō (school of engineering) submitted to the Dajōkan a few months after the submission of the Nankō plan (November 1871) includes an almost identical idea of thorough adoption of Western material culture. See Wada 2011: 89.
22. Kurasawa 1983: 293f.
23. My translation. *TDh, Shiryō 1*: 26.
24. Schwantes 1955: 157.
25. *TDh, Tsūshi 1*: 344–345.
26. Beauchamp 1976: 71, 74, and 84–85.
27. Angulo 2012, especially the useful tables comparing French and American polytechnics in the appendix (pp. 330–333).
28. Cf. Miyoshi Nobuhiro's comment to Kurahara 1987b, in *Nihon kyōikushi kenkyū* no. 6: 22–25; on 24. His comment was made in response to Kurahara Miyuki's assertion that Ministry of Education

officials did not understand Griffis's "modern" concept of "special studies" (Kurahara 1987b: 2).

29. Beauchamp 1976: 85. See also Kurahara 1987b: 4–5. The School Regulation of the Nankō (*Nankō Kisoku*) established in April 1872 listed the name of Griffis in the English division and where chemistry appeared for the first time in the curriculum of the general or preliminary course, in *TDh, Tsushi 1*: 179–184.
30. On Griffis's life until the appointment at Tokyo, see Beauchamp 1976: 7–69, Yamamoto 1987: 1–31, and Kurahara 2000.
31. Beauchamp 1976: 89, Kurahara 1987a, Gooday and Low 1998: 106, and Kurahara 2000: 150.
32. Helbig 1966: 60–61 and Yamamoto 1987: 26–27.
33. Kurahara 1987a: 24.
34. Roscoe 1868, Barker 1870, and Eliot and Storer 1871. Uchida et. al. 1990: 250 and Oki 2013: 191–198 proved that, based on Barker 1870, Griffis's lectures included extensive discussion of Avogadro's hypothesis reinterpreted by Cannizzaro and the distinction between atoms and molecules, one of the first attempts in Japan. Other references used by Griffis were Miller 1867 and Bowman 1866. These two are connected to the teaching of practical chemistry at KCL: Miller was professor of chemistry there, and Charles L. Bloxam, who edited Bowman 1866, was a RCC alumnus and a professor of practical chemistry at KCL.
35. Beauchamp 1976: 49.
36. Griffis 1924: 1195 ("I thank my Creator...") and 1196 ("certainly no other branch..."). Griffis belonged to a long tradition of natural theology in the nineteenth century. See, for example, Brooke 1991: 192–225.
37. Gooday and Low 1998: 106.
38. My translation. Japanese original cited in Tsukahara 1978a: 55–56.
39. *TTDg*, vol. 1, 338–345; and Griffis's entries in his diary in Kurahara 2005: 24 and 39. The importance of this event as a "royal affair" was discussed in Duke 2009: 154–155.
40. The introduction of the annual report of the Tokyo Kaisei Gakkō for 1875 reads: "Earlier this school consisted of divisions of three countries, but they taught languages only and had no problem at all at that time. However, when pupils progress and enter specialist departments, we need three languages and three teachers to teach each subject, the expenditure inevitably doubles or triples and we cannot afford it." My translation. *TDn*, vol. 1: 24. See Koizumi 1973: 104–109 for the problem of instructional languages at the Kaisei Gakkō.
41. *TDn*, vol. 1: 8. Materials from Osaka and Shizuoka arrived in February 1874 and "almost tripled" the number of apparatuses owned by the Tokyo Kaisei Gakkō. For the transfer of teaching materials from Osaka Rigakkō, see Shiba 2006: 49. According to

Clark, the prefecture government was generous in providing him with funding for teaching equipment (Kurahara 1997: 3–4).

42. Tanaka and Takata 1993 and Nish 1998.
43. *Meiji 6 nen 5 gatsu Monbushō Ukagai 2* (2A-9-Kō 781: 3), *Kōbunroku*. See chapter 1, notes 48 and 49.
44. My translation. Ōki's reply to Itō in March 1873, *Meiji 6 nen 5 gatsu Monbushō Ukagai 2* (2A-9-Kō781: 5), *Kōbunroku*.
45. My translation. Transcribed in Kurasawa 1973: 372–375. Also cited in Gotō 1985: 26–27.
46. Tanaka 1875, vol. 2: 47–64, on 52–64.
47. Yoshiie 1992: 79–85
48. My translation. Letter from the Kaisei Gakkō to the Ministry of Education, 4th day of the 3rd month, Meiji 6 nen (1873). "Kaigai kakkoku e kyōshi chūmon no ken," Sheets 516–524, on Sheets 516–517, *Meiji 6 nen hei* (A-7). *Monbushō ōfuku*.
49. "Senmon Gakkō on setsuritsu ni tsuki shoka kyōshi yatoiire no gi ukagai" (2A-9-kō 205), *Kōbunroku*.
50. See Beauchamp 1976: 99–108 for a standard account of this dispute.
51. Beauchamp 1976: 99; Gooday and Low 1998: 106–107.
52. Kurahara 1987b: 8–15. See also note 28 above.
53. Beauchamp 1976: 99–100.
54. Letter to Tanaka Fujimaro (acting minister of education), Ban and Tanaka (directors, *Kaisei Gakkō*), October 13, 1874. William Elliot Griffis Collection, Rutgers University Library. Other parts of this letter were cited in Beauchamp 1976: 104–105. Japanese translation in Kurahara 1987b: 13–14.
55. Sidar 1976: 15–35.
56. Helbig 1966; 59–62.
57. *Seventh Annual Report of Rutgers Scientific School, for the Year 1871* (Trenton, NJ: Murphy and Bechtel, 1871); Massachusetts Institute of Technology, *Sixth Annual Catalogue of the Officers and Students, and Programme of the Course of Instruction, 1870–71* (Boston: A.A. Kingman, 1871); Picketts 1930; *Catalogue of the Officers and Students of Columbia College, for the Year 1875–1876; being the one hundred and twenty-second since its Foundation* (Printed for the College, 1875)
58. My translation. *TDn*, vol. 1: 6–21, on 7.
59. Beauchamp 1976: 107. See also Kurahara 2005: 43.
60. Beauchamp 1976: 107. Griffis's recollection in 1885, in Griffis 1886: 12.
61. Letter from Hatakeyama Yoshinari as director of the Kaisei Gakkō to Kido Takayoshi as the head of the Ministry of Education, February 12, 1874. "Beijin Gurifisu yatoitsugi no ken," Sheets 487–488, on sheet 487, Meiji 7 nen kō (A-9), *Monbushō ōfuku*.

62. On Murray's appointment as superintendent of the Ministry of Education, see Gotō 1985 and Gotō Sumio 1986. On Hatakeyama's role as Murray's interpreter and advisor during his office in Japan, see Murray's own account in Murray 1917: 113, as well as Yoshiie 1998: 137–139, 150, 151, and 190. However, Yoshiie did not discuss Murray's involvement in the teaching at the Tokyo Kaisei Gakkō in detail.
63. Mrs. Williamson's journal, February 20, 1874. This information comes from an excerpt (made by J. Harris) from the journal of Williamson's wife Emma Catherine, the original of which is now lost. Harris and Brock 1974: 125 (n. 105).
64. Letter from Tanaka Fujimaro to the Tokyo Kaisei Gakkō, August 3, 1874. "Eijin Akkinson. Robaru Sumesu Yatoiire no ken," Sheets 93–95, on sheet 93, Meiji 7 nen kō (A-9). *Monbushō ōfuku.*
65. *Shiryō oyatoi gaikokujin*: 209, 292–293, 341–342, and 344–345.
66. Letter from Murray to Griffis, January 24, 1874. William Elliot Griffis Collection, Rutgers University Library.
67. Smith 1897: 582.
68. See, for example, Duke 2009: 155.
69. Murray 1917: 108, 111, and 113. For Hatakeyama's patriotic sentiments and his sense of belonging to the ruling class, see Inuzuka 1987: 214–231.
70. There are several published sources for the information of chemistry teaching at Atkinson's department. Annual reports of the Tokyo Kaisei Gakkō and Tokyo University include reports by Atkinson and Jewett. See *TDn*, vol. 1: 30–31, 48–50, 67–68, 89–91, 120, and 154–155. University calendars contain both syllabuses and examination papers of Atkinson's and Jewett's lectures in English and Japanese. *Jpn. Calendar Tokyo Univ. Law, Sci., and Lit. 1880–1881*: 50–59 and 208–217; and *Eng. Calendar Tokyo Univ. Law, Sci., and Lit. 1880–1881*: 37–42 and 114–118. *A Classified List of the English Books in the Tokio-Kaisei-Gakko* (1875) is a useful source to search textbooks and references used there. *Course of Laboratory Instruction in Chemistry. 3rd Year Class. 1st Term* (University of Tokyo, n. d.) is based on Thorpe 1873. See also Tsukahara 1978a: 93–95.
71. My translation. Sakurai 1940: 9. The subjects taught but not mentioned by Sakurai include mineralogy and geology, physics and physical laboratory, and mining. Mathematics, drawing, and natural history were taught in the preliminary course. *TDn*, vol. 1: 11 and 12.
72. Starting with nine students in 1874, Atkinson supervised the laboratory work of 18 to 23 chemistry students per year with Masaki until 1876 and thereafter on his own until 1877. Geology and mining students joined the chemical laboratory starting in 1878, but this extra work was offset by the appointment of Jewett in 1877, and the number of laboratory students under Atkinson's care remained

at around 20 until his departure in 1881, in comparison to 8 to 15 under Jewett. Relevant data was taken from *TDn*.

73. Works on his life include Ueno 1968, Tomozawa 1965, and Shiokawa 1977.
74. Apart from chemistry and chemical laboratory, Atkinson took mathematics, geology, mineralogy, physics, botany, physical laboratory, and political economy at UCL. The curriculum of the RSM was focused on the teaching of the basic principles of the sciences underlying mining. There Atkinson studied physics, mechanical drawing, mineralogy, geology, applied mechanics, and metallurgy at the RSM. He did not take engineering subjects, applied chemistry, or chemical technology. See UCL Student Register, UCL Records Office; College of Chemistry Cash Account, Commencing 29th July 1844 (C6a.561) and Examination Returns of the RSM (D9/1 876), Imperial College Archives.
75. Atkinson 1872. See, for example, Brock 1967: 15 for the debates on the atomic theory involving Williamson, Frankland, Odling, and Brodie in 1869. See Fujii 1975 for Atkinson's polemic together with Japanese chemists' views on atomism.
76. *Kotō Bunjirō hikki Atokinson kōgi nōto*, 7 vols., in the collections of the Tokyo Daigaku Sōgō Kenkyū Hakubutsukan. As Kotō majored in geology, this source does not cover Atkinson's lectures on chemical technology.
77. See, for example, Atkinson's extensive discussion of aromatic compounds in Kotō's notebook on organic chemistry, vol. 2: 51–118 and vol. 3: 1–27.
78. Atkinson and Jewett's choice of textbooks corroborates this point and strongly reflects contemporary practices in English chemistry teaching while maintaining continuity with that of his predecessor. Atkinson combined the textbook already used by Griffis, Roscoe 1868, with such classics as Fresenius' *Qualitative Chemical Analysis* and *Quantitative Chemical Analysis* and the latest publications including Schorlemmer 1874 and Thorpe and Muir 1874, both published in London in 1874. Atkinson's choices thus had a large share of publications from Owens College Manchester, including Jones 1872 in addition to Roscoe, Schorlemmer, Thorpe, and Muir. Under Roscoe's tutelage, Owens College emerged in the 1860s and 1870s as a center of chemistry teaching in England that rivaled UCL and RSM. See Bud and Roberts 1984: 83.
79. See, for example, Jewett's notebook "Lectures on Organic Chemistry by Prof H. Hübner. Göttingen 1874–75" (acc. 2001/94). His laboratory notebook from his student days records thorough training in qualitative and quantitative analysis (both inorganic and organic) including combustion analysis, albeit, interestingly, without the *kaliapparat*. "F. F. Jewett: Notes on Lectures and Chemical

Experiments, Yale University and Göttingen" (acc. 2001/94): 135, both in Oberlin College Archives. Also, in his report to the Tokyo Kaisei Gakkō, Jewett discussed how the "carelessness" and "filthiness" of some students had failed them in learning the detection of alkaloid poison. *TDn*, vol. 1: 68, as cited in Watanabe 1973: 223.

80. *TDn*, vol. 1: 55.
81. Bud and Roberts 1984: 75–79 (UCL's chemistry lecture course) and 81–86 (chemistry lecture course at Owens College Manchester). Atkinson did not indicate any textbook for this course, but he may well have drawn on Ronalds, Richardson, and Watts 1855–1867, which is mentioned in *A Classified List of the English Books in the Tokio-Kaisei-Gakko.* Interestingly, Ronalds was a superior of Divers while the latter was an assistant at Queen's College Galway, and it may well be the case that Divers, who came to Japan earlier, suggested this title to Atkinson. See Kikuchi 2012: 294–5 and 299–300.
82. My translation. *TDn*, vol. 1: 29.
83. Note 78 above.
84. *TDn,* vol. 1: 49, 67–68 and 89–90.
85. See, for example, Atkinson 1877, Atkinson 1878, Atkinson 1879a, Kuhara 1879a, Kuhara 1879b, and Kōga 1880. Atkinson's impact on Japanese water analysis in the early Meiji period is analyzed in Shiokawa 1977 and 1978.
86. The laboratory notebook of Alfred Chaston Chapman, Graham's student and later assistant in his consulting practice, taken during the sessions 1886–7 and 1887–8 at UCL shows the details of Graham's laboratory teaching characterized here. Chemical Notebook of Alfred Chaston Chapman (MS140), Leeds University Library Special Collections. Though it is not a contemporary source of Graham's laboratory teaching in the early 1870s, it does indicate Graham's laboratory practices accumulated through two decades of his teaching at UCL and therefore, in my opinion, could be used to infer his earlier teaching practice that was later developed into Graham's own course of chemical technology at UCL in the late 1870s and 1880s.
87. Robert William Atkinson, Certificate of a Candidate for Election, The Chemical Society, June 18, 1872 (balloted March 7, 1872), Royal Society of Chemistry.
88. Russell 1996: 362. Atkinson mentioned Frankland's water analysis in Atkinson, "Tokyo-fuka yōsui shiken setsu," p. 26 *et passim.* Interestingly, Atkinson used in his teaching at Tokyo the textbook of Frankland's archrival, J. Alfred Wanklyn, Wanklyn and Chapman 1868.
89. My translation, *TDn*, vol. 1: 89.
90. My translation, from Atkinson 1879b: 139–140. See also Graham's original in Graham 1873–1874: 190.
91. *TDn*, vol. 1: 89f and 154f.

92. Takamatsu 1878, Isono 1879, and Nakazawa 1879.
93. Atkinson 1881a; abridged version, Atkinson 1881b; and Japanese translation, Atkinson 1881c. See Tomozawa 1965 for Atkinson's research on sake brewing and the response of Japanese brewers to Atkinson's research. See also Morris-Suzuki 1994: 109. Neither Tomozawa nor Morris-Suzuki discussed pedagogical origins of Atkinson's research on sake.
94. See Nishizawa 1923: 579 for Takamatsu's support of Atkinson's research on sake. Atkinson acknowledges Nakazawa's support in Atkinson 1881a: iii.
95. Tomozawa 1965: 117–118.
96. Graham 1873–1874: 333–334.
97. Morris-Suzuki 1995.
98. Atkinson, Takamatsu, and Isono all acknowledged manufacturers' support not only as suppliers of samples but also as informants. See Takamatsu 1878: 2, 13, 39 and 42; Isono 1879: 19 and 25; and Atkinson 1881a: iii.
99. My translation, *TDn*, vol. 1:90.
100. *Jitsugaku* literally means "real learning," "practical learning" or "learning with substances" and was used polemically by a variety of scholars and thinkers in the Tokugawa period to criticize the "futility" of the abstract theories of their rivals, theories they alleged were not based on actual experiences. Here I follow the definition of *jitsugaku* by Sugimoto as indigenous empirical learning for utilizing natural resources for the benefit of people (*riyō kōsei no gaku*). See Sugimoto 1962.
101. Takamatsu 1878: page after p. 47.
102. Isono 1879: 12 and 16f.
103. Sugimoto 1967: 148–204, on pp. 161–165.
104. See the frontispiece in Atkinson 1881a. Its Japanese translation, Atkinson 1881c, also carried the illustration without a reference.
105. Atkinson 1881a: VIII. See also Atkinson 1881a: 63.
106. See, for example, Lissa Roberts's contribution on science education and citizenship in late eighteenth-century Netherlands, in Heering and Wittje 2011: 71–95.
107. Herzfeld 2002. "Improvement" as a colonial discourse mobilized both by British overlords and Indians is analyzed in Watt and Mann 2011. I am grateful to Aarti Kawlra and Ethan Mark for their insights, and to Aarti Kawlra for the references.
108. Atkinson 1881a: 63–64.
109. Wada 2012 and Kikuchi 2012: 297–302.
110. The most common topics in Divers's examination papers on applied or technological chemistry were cement, glass (both as early as in the *ICET Calendar 1876–1877*: LXXXVII), petroleum (its earliest appearance in *ICET Calendar 1878–1879*: CXII) and soap (its

earliest appearance in *ICET Calendar 1879–1880*: XCVIII-XCIX) together with coal gas and artificial dye manufactures, but the alkali industry entered Divers's examination rather late, first in the *ICET Calendar 1883–1884*: LXXVI. The only traditional Japanese manufacture featured in Divers's examinations was sake brewing, which appeared only once in the *ICET Calendar 1883–1884*: LXXVII after the publication of Atkinson's article on this topic in 1881.

111. *Shiryō oyatoi gaikokujin*: 209 and 292–293. Atkinson's salary was 350 yen per month in comparison with Jewett's 300 yen, which was increased to 350 in 1880. Hired teachers in the early Meiji period mainly received salaries in Spanish dollars, and 1 yen was worth around 1 Spanish dollar. See Uemura 2008: 5.
112. For Divers's greater power as professor than lecturers in curricular matters, see Kikuchi 2012: 300.
113. *Shiryō oyatoi gaikokujin*: 315. Divers's first contract with the Ministry of Public Works was for five years; he then concluded a new one with indefinite term. Records on his salary at the Imperial College of Engineering are ambiguous, but one of them recorded a much higher salary (600 yen per month) than that of Atkinson or Jewett even before Divers assumed the office of principal, to which was added a salary as the ministry's analyst.
114. Haga 1912: 15.
115. Gooday 1991: 87–88 and Kakihara 1996.
116. See, for example, Horikoshi 1973 (which included the Tokyo Kaisei Gakkō building) and Kondō 1999.
117. *Takamatsu Toyokichi den*.
118. My translation, from Nishizawa 1923: 499. Takamatsu referred to the Tokyo Kaisei Gakkō by its older name, Nankō, and to the Imperial College of Engineering, Tokyo, by its abbreviated name, Kōgakkō.
119. My translation, from Nishizawa 1923: 579.
120. For plans of the Imperial College of Engineering, see Brock 1981: 234f.
121. Shinoda 1944. Cited in Tsukahara 1978a: 91–92.
122. Kikuchi 1999: 105. See also Kamatani 1988: 35 and 43.
123. Kamatani 1988: 33–49 (Geological Survey) and 80–112 (Industrial Research Laboratory). Morris-Suzuki 1995: 110.
124. Kyū Kōbu Daigakkō Shiryō Hensankai 1931: 349–352
125. Wada 2012: 58, 60 (table 3) and 69.
126. There are numerous works on Takamine. For a well-researched biography, see Iinuma and Sugano 2000.
127. Mills and Takamine 1883.
128. Wada 2012: 59 (table 59). I supplemented his data with those from my ongoing unpublished prosopographical research on the careers of Imperial College of Engineering graduates such as:

Tanabe Hidenosuke (grad 1880), Kyushu Sekiyu Seizō (Petroleum Manufacture); Imai Zenshichi (grad. 1880), Keijō Copper Mine in Korea; and Shizuki Iwaichirō (grad. 1884), Nihon Seimi Seizo (chemical manufacture). On the methods and references used for this investigation, see Kikuchi 1999.

129. Shiohara 1926: 27–28; Iinuma and Sugano 2000: 55–56.
130. Iinuma and Sugano 2000: 110. It mistakenly mentions Hida as a graduate of the Imperial College of Engineering. For correct information, see Kikuchi 1999: 105.

3 The Making of Japanese Chemists in Japan, Britain, and the United States

1. Amano 2009, vol. 1: 332–334. See also Nakayama 1978: 102–104.
2. For an exemplary discussion on the high-brow "samurai" characteristics of Japanese engineering, see Muramatsu 1954, quoted in Wada 2012: 58 and 73. On comparable discussions of Japanese physics in the nineteenth and twentieth centuries, see Koizumi 1973, Koizumi 1975, and Low 2005: 6 and 15.
3. Koizumi 1973: 141, 143, and 151–2; Koizumi 1975: 50, 51, and 53; and Dore 1965: 97–98. For the *kōshinsei* system, see Karasawa 1990.
4. Karasawa 1990: 5–6 and 41–42.
5. Dore 1965: 33–67.
6. Koizumi 1973: 205; Koizumi 1975: 77.
7. Sugiura is today best known as the ethics tutor of Emperor Hirohito when the latter was crown prince. See, for example, Wetzler 1998: 86–113.
8. Ōmachi and Ikari 1986: 107–108, 119, and 129.
9. Kozai 1911.
10. Tsukahara 1978a: 22–31, Gotō 1980a, and Gotō 1980b. Though not widely circulated, the 2005 exhibition booklet on Kuhara and his family, *Tsuyama han-I Kuhara-ke to kagakusha Kuhara Mitsuru*, is a useful guide on his family background and archival sources.
11. Kuhara 1877 reveals his early interest in the sociomedical effects of chemicals (most probably coming from his family background) before turning his attention to industrial chemistry under the guidance of Atkinson.
12. Karasawa 1990: 3.
13. Sakanoue 1979: 3 and Sakanoue 1997: 157–158.
14. Sakurai 1940: 2. On the life of this teacher, Percival Osborn, see Imai 1994.
15. Sakurai 1940: 7–8.
16. *Takamatsu Toyokichi den*: 4.

17. *Takamatsu Toyokichi den*: 5–6. For the commoners' education at *terakoya*, see Dore 1965: 252–292.
18. Dore 1965: 49.
19. My translation. *Takamatsu Toyokichi den*: 403–404.
20. See Takamatsu's recollection in Nishizawa 1923: 497 and 579.
21. *Takamatsu Toyokichi den*: 302.
22. The report of the class examination in July 1875 is cited in Hozumi 1988: 112. See also Sakurai's certificate from the Tokyo Kaisei Gakkō dated August 19, 1876, in Sakurai Jōji Papers: G-1.
23. "Rigaku hakushi Matsui Naokichi kun shōden" (1888): 105.
24. Sakurai 1940: 3. See also Nishizawa 1923: 81.
25. Karasawa 1990: 400–404, on 400.
26. Karasawa 1990: 401 and 402. See also Tsukahara 1978a: 70–71 and *Tsuyama han-i Kuhara-ke to kagakusha Kuhara Mitsuru* (2005): 15.
27. Letter from Kuhara to his father, March 11, 1878. In Tsukahara 1978b: 11. See also Gotō 1980a: 20.
28. *TDn*, vol. 1: 55.
29. Gotō 1980a: 20; *Takamatsu Toyokichi den*: 2.
30. My translation, from *TDn*, vol. 1: 90.
31. Takamatsu 1878: 1. *Takamatsu Toyokichi* den: 195–196.
32. Takamatsu mentions Mr. Murakawa's *oshiroi* (Japanese white cosmetics) works in Kyoto, Mr. Murata in Tokyo, Mr. Tateyama of Kyoto, Mr. Namikawa of Kyoto. See Takamatsu 1878: 2, 7, 13, 39, and 42.
33. Takamatsu 1878: 12.
34. Takayama Jintarō (grad. 1878) and Ota Kenjirō (grad. 1879) of the Chemistry Department, and Watanabe Wataru (grad. 1879) of the Mining/Metallurgy Department of Tokyo University. See Takamatsu 1878: 5, 34 and 43.
35. Kuhara 1879c. The other two articles in this series are Atkinson 1880 and Moriya 1881.
36. Kuhara 1879c: 22.
37. Kuhara 1879c: 25. As Catherine Jackson argues, these preparations for analytical purposes, together with elemental analysis, became the backbone of the methodology of structural organic chemistry by the middle of the century and an integral part of organic chemists' concept of "synthesis" by the end of the nineteenth century. Jackson: 2008a: 116–161.
38. See Ōmachi and Ikari 1986: 117, Hozumi 1988: 106–107, and Karasawa 1990: 78–97. It is not clear whether Matsu and Sakurai were involved in this movement.
39. For a recent historical treatment of Columbia University and its constituent schools, see McCaughey 2004: 152–155.
40. See, for example *Columbia College Catalogue 1875–1876*: 73 and 78. For the Tokyo curriculum, see chapter 2, note 71.
41. *Columbia College Catalogue 1875–1876*: 93 and 94.
42. For studies of Chandler's life using the Charles F. Chandler Papers now deposited in the Columbia University Rare Book and Manuscript

Library, see Larson 1950 and Rossiter 1977. See also McCaughey 2004: 153.

43. "Lectures of Professor C. F. Chandler on "Industrial Chemistry," given in the first semester 1902–1903 to the School of Applied Science, and written by H. H. Higbie from his lecture notes," the Rare Book and Manuscript Library, Columbia University Libraries.
44. See note 40 above.
45. Letter from Chandler to Mekata, September 15, 1876. Japanese translation in sheets 387–388, Meiji 9-nen otsu (A-16), *Monbushō Ōfuku*. Matsui did well also in analytical and descriptive geometry, botany and zoology, general chemistry, crystallography, and stoichiometry, but he did not take drawing.
46. Letter from Mekata to the head of the Ministry of Education, October 21, 1878, in Meiji 11-nen otsu (A-24), sheet 569, *Monbushō Ōfuku*.
47. N. Matsui, "Memoir on Enamelling of Iron Wares" (MINES: Ms 1877), Columbia University Libraries.
48. *Columbia College Catalogue 1877–1878*: 111.
49. Crookes and Röhrig 1869, vol. 2: 675–684; Ure 1856, vol. 1: 651. In his essay Matsui only referred to the latter without bibliographic details. One example of Matsui's almost verbatim quotation of Crookes and Röhrig 1869 is:

> The mass for the first coating consists chiefly of silica, boracic acid, alumina, and alkalies; the enamel or second coating contains the same ingredients ["components" in Crookes and Röhrig 1869, vol. 2: 676] but a larger proportion ["amount" in Crookes and Röhrig, ibid.] of alkali, which makes it ["the mass" in Crookes and Röhrig, ibid.] more easily fusible, and also some oxide of tin (SnO_2 [no parenthesis with a chemical formula in Crookes and Röhrig, ibid.]), rendering it opaque and white. With regard to ["Concerning" in Crookes and Röhrig, ibid.] their chemical composition both mixtures are variable within a certain limit. Some enamel also contains lead; it then may be fixed to the iron at a lower temperature, and therefore, more cheaply, and it is also more durable; but it is less white in appearance and it is liable to be injurious when used for cooking vessels.
>
> Eulenberg ascertained this enamel to contain –
>
> | | |
> |---|---|
> | Silica | 43.38 |
> | Oxide of lead | 39.12 |
> | Phosphoric acid. | 3.51 |
> | Phosphate of lime | 2.61 |
> | | 88.62 |
>
> The production of enamels free from lead has lately been aimed at (Matsui, "Memoir on Enamelling of Iron Wares," Sheets 6–7).

50. James Atkins Noyes, "Memoirs On Enamels for Iron Ware" (MINES: Ms 1877), Columbia University Libraries.
51. Noyes, "Memoirs On Enamels for Iron Ware," Sheets 2–6.
52. Rossiter 1977: 224.
53. See note 41 above. The other two topics were "The Best Method for Separating and Determining Nickel, Cobalt, and Zinc" and "The Separation and Determination of Iron, Chromium, and Titanium in Chromic Iron."
54. Matsui 1880.
55. Matsui 1880: 315–316.
56. Matsui 1880: 331.
57. For a useful near-contemporary account of the Arita porcelain to contextualize Matsui's paper, see Brinkley 1901: 64–74, on pp. 69 and 73.
58. Sakurai 1880a.
59. Sakurai 1880b.
60. Williamson 1865: iii. This point is made in Harris and Brock 1974: 115f, and Rocke 2004: 340.
61. "Notes of Lectures delivered by Professor Williamson, Ph.D., F.R.S. at University College, London, on Chemistry, during the Session 1877–1878" (MSS 2675–6). Lecture Notes Collection of William Dobinson Halliburton, Wellcome Library. This source is mentioned in Rocke 2004.
62. Harris and Brock 1974: 115.
63. See also Gooday 1989, chapter 4: 3–6 and 8.
64. "Notes of Lectures delivered by Professor G. Carey Foster, B.A., F.R.S. at University College, London, on Experimental Physics, during the Session 1877–1878" (MSS 2677–8), Lecture Notes Collection of William Dobinson Halliburton, Wellcome Library. Halliburton took lecture notes of Division A and Division B in the same notebooks in reverse directions. For Foster's physical laboratory teaching at UCL and his origin as a chemist, see Gooday 1989, chapter 4, and Brock 1998.
65. Bud and Roberts 1984: 79–81.
66. UCL Senate Minutes, October 13, 1874, and November 20, 1874; UCL Records Office. UCL Council Minutes, November 7, 1874, November 21, 1874, and December 5, 1874; UCL Records Office.
67. Sakurai took chemistry from 1876–1877 until 1879–1880, chemistry exercise class in 1876–1877 and analytical chemistry from 1876–1877 until 1880–1881. Among physics-related subjects, Sakurai took physics Division B (1877–1878), physics (1878–1879, 1879–1880, 1880–1881) and physics exercise class (1878–1879). See Arts A 1876–77, Arts A 1877–78; Arts B 1877–78; Arts B 1878–79; Arts C 1878–79; Arts B 1879–80; and Arts A 1880–1881, UCL Student Registers, UCL Records office. In his autobiography,

Sakurai regarded Williamson, Foster, and Lodge his principal teachers at UCL. See Sakurai 1940: 12.

68. My usage of the term "physicalist" follows John Servos, who categorized its tradition in nineteenth-century chemistry as: (1) the development of physical instruments and their application to the study of chemical composition, (2) research on the relationship between physical properties and chemical composition, and (3) the study of the physical principles that governed the processes of chemical changes. Servos 1990: 11.
69. Letter from Sakurai to Alice Maude Fison, January 12, 1927. Collection of notes and correspondence about Alexander William Williamson. UCL Library Services Special Collection.
70. Sakurai 1880b: 661.
71. In the abstract version of Sakurai's paper, he acknowledged that the choice of methylene iodide as the starter reactant was made on the suggestion of Williamson. Sakurai 1880a: 504.
72. Letter from A. W. Williamson to Sakurai, August 20, 1880. Sakurai Jōji Correspondence, 1–68. Transcribed in Kikuchi 2004: 258.
73. Letter from A. W. Williamson to Sakurai, September 19, 1882. Sakurai Jōji Correspondence: 1–69. Transcribed in Kikuchi 2004: 258–259.
74. Morrell 1972 and Rocke 2003: 92 and 100–107.
75. Harris and Brock 1974: 100–108.
76. *Account of the New North Wing and Recent Additions to University College, London* [...] (1881): Plate 11.
77. Letter from Sakurai to Fison, January 12, 1927 (cf. note 69 above). See Barff 1878–1879 for Barff's method.
78. After the completion of the 1881 Laboratory, the Birkbeck Laboratory was handed over to Charles Graham for his laboratory course on chemical technology. See *An Account of the New North Wing and Recent Additions to University College, London* [...] (1881): 11–12. Cited in Donnelly 1987: 123.
79. Robins 1887: 95.
80. Arts B 1878–1879; and Arts B 1879–80, UCL Student Registers, UCL Records Office.
81. See Joji Sakurai, Certificate of a Candidate for Election, the Chemical Society, April 24, 1879 (balloted June 19, 1879), Royal Society of Chemistry.
82. See Bud and Roberts 1984: 121, 130 and 135–136.
83. See Sakurai's certificate of election to the membership of the Society of Arts, Sakurai Jōji Papers: G-4, and the Membership Record, Royal Society of Arts Archives.
84. Sakurai 1899: 17.
85. My translation, from Sakurai 1899: 17–18.
86. Sakurai 1899: 10–11.

87. Kikuchi 2004: 239. This process continued well after Sakurai returned to Tokyo in 1881. Sakurai practiced British-style horse riding, and together with his wife he entered a weekly dance club based in *Rokumeikan*. See the membership card of the Dance Society (*Bugakukai*) for Sakurai and his wife San (or Sanko) as well as his certificate for equestrian skill (*bajutsu renshūsha shō*). Sakurai Jōji Papers: G-10.
88. O. L. Brady, "A Short History of the Chemical and Physical Society" (October 1924), unpublished typescript, UCL Chemistry Department.
89. A similarity between this title and Takamatsu 1878 suggests that Sakurai was informed by Takamatsu in preparing for this presentation. In fact, there is evidence that Takamatsu went to London several times during his vacations and that Takamatsu met Sakurai during his overseas study in Manchester between 1879 and 1881. See Sakurai 1932.
90. Roscoe 1878. According to Peter Morris, spy windows originated in Hermann Kolbe's new laboratory at Marburg in 1863, were taken up by Roscoe at Manchester, and spread to Newcastle but never seem to have spread to other German universities, most conspicuously not to Heidelberg where Roscoe studied with Bunsen. See "The Director's Residence and the Spy Window," in Morris, *Crucible of Science: The History of the Chemical Laboratory* (London: Reaktion Press, forthcoming). I am grateful to Peter Morris for sharing his manuscript.
91. The Owens College, Chemical Department, Reports (DCH/1/2/3/1). University of Manchester Archives, John Rylands University Library of Manchester. This source is hereinafter referred to as "Owens Chemical Department Reports."
92. Owens Chemical Department Reports: 133 and 162.
93. Roscoe 1887: 45–52. See also Bud and Roberts 1984: 84–85 and 213–214.
94. Takamatsu and Smith 1880, Smith and Takamatsu 1881a, 1881b, and 1882.
95. Bud and Roberts 1984: 85 and Takamatsu and Smith 1880: 592. They reported that the "doubts which are entertained respecting the very existence of pentathionic acid led us, at Dr. Roscoe's suggestion, to undertake the investigation of which we now give the results."
96. Toyokichi Takamatsu, Certificate of a Candidate for Election, the Chemical Society, January 20, 1881 (balloted February 17, 1881), Royal Society of Chemistry. The recommenders "from personal knowledge" were five Mancunian chemists: Roscoe, Schorlemmer, Smith, Peter Phillips Bedson, and William Carleton Williams; in addition to Sakurai, who stayed in London until August 1881. See Bud and Roberts 1984: 85 for the phrase "scientific chemistry."
97. After Takamatsu stayed briefly in Manchester on his way back from Germany to Japan in July 1882, he was escorted by Smith and his

wife to the port of Liverpool and seen off by them. *Takamatsu Toyokichi den*: 29.

98. *Owens College Calendar 1881–1882*: 198.
99. See, e.g., Bud and Roberts 1984: 81 for the aims of Owens College Manchester as perceived by its founders. For the syllabus and examination papers of the Owens Chemistry Department during Roscoe's professorship and before the arrival of Smith, see, e.g., *Owens College Calendar 1876–1877*: 52–55 and clxxxix–cxciii and Roscoe's testimony to the Devonshire Commission on March 31, 1871, q. 7367.
100. For Smith's course, see Donnelly 1987: 127–154 and Donnelly 1997: 131–136.
101. See the obituaries of Smith in *Journal of the Society of Chemical Industry: Reviews* 39 (1920): 191–192 and in *The Chemical Age* 2 (1920): 504.
102. As is exemplified by the second section "Construction and Description of Plant, etc" (drawing materials requisite in this examination) of Smith's examination papers in *Owens College Calendar 1881–1882*: clxxii-clxxv. A typical question in 1880–1 is: "By means of description and sketch in vertical section, show how the lead of a sulphuric acid chamber sides and top is secured to the crown-trees and top-joists."
103. My translation. *Takamatsu Toyokichi den*: 26–27.
104. Roscoe's statement on June 18, 1868, in Samuelson Committee, q. 5618.
105. *Takamatsu Toyokichi den*: 27.
106. According to Takamatsu 1882, one of Takamatsu's topics during his work at Hofmann's laboratory was to learn how to synthesize indigo. This article does not show any hint that his work was on an industrial or mass scale.
107. My translation, from Hofmann's inaugural address as rector of the University of Berlin in 1880. Cited in Scholz 1992: 225.
108. Johnson 1992: 171.
109. Roscoe 1881.
110. Donnelly 1996: 784–785. See also Russell, Coley, and Roberts 1977: 61–65.
111. My translation, from the letter from Kuhara to his father, April 14, 1879, in Tsukahara 1978b: 14–15, on 14.
112. Deposited in the Special Collections, The Milton S. Eisenhower Library, JHU.
113. I owe this point to James Stimpert, university archivist of the Sheridan Libraries, JHU.
114. *JHUC* no. 4 (April 1880): 52. For an early history of JHU, see Hawkins 2002.

115. My translation, from Kuhara's letter to his parents, October 1879, in Tsukahara 1978b: 22–23, on p. 22. See also his letters to his parents, September 27, 1879, and November 5, 1879, in idem, 20 and 23–24.
116. *JHUC* no. 1 (December 1879): 3 and no. 2 (January 1880): 10 and 12. He was officially enrolled in "Laboratory Work" for 1879–80 (first half year) and 1880–81 (the whole year). See idem, no. 2 (January 1880): 10; no. 7 (December 1880): 74; and no. 10 (April 1881): 122. He also took a minor course in German in the first half-year of 1879–80, the latter chosen with advice from Remsen. For his advice, see Kuhara's letter to his parents, December 27, 1979, in Tsukahara 1978b: 24.
117. Hannaway 1976: 146 and 158. The latter quoted in Warner 2008: 51.
118. Warner 2008: 50–52.
119. Warner 2008: 58–59.
120. *JHUC* no. 5 (May 1880): 59. The earliest official usage of the term "Journal Meetings" appeared in *JHUC* vol. 8, no. 74 (July, 1889): 86.
121. Letters from Kuhara to his parents, February 18, 1880, in Tsukahara 1978b: 25–26; and March 11, 1880, in idem, p. 27.
122. Letters from Kuhara to his parents, January 28, 1880, in Tsukahara 1978b: 25.
123. Warner 2008: 51. Remsen's most famous student, William A. Noyes considered Remsen's "daily visit" the "most important and vital part of his instruction" (Noyes 1927: 245).
124. My translation, from Kuhara's letter to his parents, December 27, 1979 (see note 117 above).
125. Warner 2008: 51.
126. Kuhara 1879–1880. I also consulted Gotō 1980a: 23, 25–26 for the analysis of Kuhara's papers at JHU, but its usefulness is diminished by the fact that Gotō did not consider the wider research and pedagogical contexts of JHU.
127. One of Morse's obituaries written by W. H. Howell, a student in the same period as Kuhara, testified to Morse's very rigorous analytical training that typically consisted of quantitative analysis of metals. To *ACJ* Morse also contributed articles on this topic between 1880 and 1900 before publishing a series of papers on the electrolytic method of the measurement of osmotic pressures, which garnered him recognition as a scientist. See Remsen 1923.
128. Remsen and Kuhara 1880–1881, Kuhara 1881–1882, and Remsen and Kuhara 1881–1882. Kuhara's Ph.D. was conferred in absentia in January 1882 after his departure from San Francisco for Yokohama in November 1881. See *JHUC* no. 16 (July 1882): 236 and two letters from Kuhara to his father, November 29, 1881, and January 27, 1882, in Tsukahara 1978b: 42–43.

129. On Remsen's research program, see Tarbell and Tarbell 1986 in addition to Hannaway 1976 and Warner 2008.
130. Remsen and Burney 1880–1881.
131. Remsen and Fahlberg 1879–1880: 438. See also Warner 2008: 52 and 57. Warner cited Fahlberg's own testimony that the method of preparing benzoic sulfinide by oxidation is entirely his.
132. Remsen and Fahlberg 1879–1880: 437–438.
133. Kuhara 1881–1882: 26–27.
134. Letter from Kuhara to his parents, October 24, 1880; letter from Kuhara to his father, July 10, 1881, in Tsukahara 1978b: 30–31, on 30, and 37–38, on 38, respectively. He spent the summer of 1881 to study at Yale's mineralogical museum with James Dwight Dana to prepare for the Ph.D. examination, for which he chose mineralogy as a minor.

4 Defining Scientific and Technological Education in Chemistry in Japan, 1880–1886

1. Allen 1962: 47–54 and Jansen 1989: 614–617 (Matsukata's fiscal policy) and 647f. (introduction of the cabinet system).
2. It is difficult to identify the dates of all their appointments with primary sources of the same type. To simplify matters, I adopted the year when they first entered the staff roster of published university calendars. When available, I consulted both English and Japanese versions to ascertain the translation of titles and subjects. See *Jpn. Calendar Tokyo Univ. Law, Sci., and Lit. 1880–1881*: 6; *Eng. Calendar Tokyo Univ. Law, Sci., and Lit. 1880–1881*: 6; *Jpn. Calendar Tokyo Univ. Law, Sci., and Lit. 1881–1882*: 171–172; *Jpn. Calendar Tokyo Univ. Law, Sci., and Lit. 1882–1883*: 204–205; and *Jpn. Calendar Tokyo Univ. Law, Sci., and Lit. 1882–1883*: 210–211.
3. My translation. "Rigakubu Kyōju Atokinson, Howittoman Manki Yatoidome Wakuneru Yatoiire no gi Ukagai" (2A-10-kō 3063: 31), *Kōbunroku*. See also "Doitsujin Waguneru Yatoiire no ken," Sheet 96, *Meiji 14 nen kō* (A34), *Monbushō ōfuku*.
4. My translation, in Sakurai 1940: 17.
5. Bartholomew 1989: 263.
6. Sheets 364–390, *Meiji 9 nen otsu* (A-16); Sheets 456–492, *Meiji 10 nen otsu* (A-19); Sheets 312–316 and 568–571, *Meiji 11 nen otsu* (A-24); sheets 230–283 and 292–297, *Meiji 12 nen kō* (A-27); and Sheets 592–596, *Meiji 14 nen otsu* (A-35); all in *Monbushō ōfuku*. "Kaigai ryūgakusei kichō no gi jōshin" (2A-10-Kō 2665: 0211), *Kōbunroku*.
7. Sheets 255, 256 and 269, *Meiji 12 nen kō* (A-27), *Monbushō ōfuku*.
8. Sheets 294 and 295, *Meiji 12 nen kō* (A-27), *Monbushō ōfuku*. Other records of Sakurai's overseas study includes sheets 470, 477 and 482, *Meiji 10 nen otsu* (A-19), *Monbushō ōfuku*.

9. When the Ministry of Education notified Katō of the return of three Japanese students from Britain, including Sakurai, to Japan in September 1881, Sakurai was mentioned at the top of the list as a "graduate of the Department of Chemistry" with two Glasgow BSc holders in Engineering, Masuda Reisaku and Taniguchi Naosada. See "Eikoku Ryugakusei Sakurai Joji hoka sanmei sotsugyo kicho no ken," Sheets 576 and 577, *Meiji 14 nen otsu* (A-35), *Monbushō ōfuku*.
10. Sheet 595 [28 July 1881], Meiji 14 nen otsu (A-35), *Monbushō ōfuku*.
11. *TDn*, vol. 2: 46–47.
12. *Ibid*. See also *TDn*, vol. 2: 149.
13. Matsui 1882. For the date of his lecture, see "Honkai kiji," *Tkk* 3 (1882): 1–15, on 3–4. On chemical atomism, see Rocke 1984.
14. For the history of the Tokyo Chemical Society and its successor society, the Chemical Society of Japan, see Nihon Kagakukai 1978, which is comprehensive but celebratory as a society publication; and Hirota 1988, a well-researched and more opinionated work on the society.
15. Masanori Kaji, "Chemical Classification and the Response to the Periodic Law of Elements in Japan in the 19th Century," paper read at the 7th International Conference on History of Chemistry, Sopron, Hungary, August 4, 2009; idem, "Chemical Classification and the Response to the Periodic Law of Elements in Japan in the Nineteenth and Early Twentieth Centuries", the 7th meeting of Science & Technology in the European Periphery, Galway, Ireland, June 19, 2010; idem, "Chemical Classification and the Response to the Periodic Law of Elements in Japan in the Nineteenth and Early Twentieth Centuries", the 4th International Conference on the European Society for the History of Science, Barcelona, Spain, November 18, 2010; idem, "Classification of Elements and the Responses to the Periodic Law in Japan up to the 1920s", the 24th International Congress of History of Science, Technology and Medicine, Manchester, United Kingdom, July 25, 2013. I am grateful to Masanori Kaji for sharing his manuscripts. An enlarged version of his presentation drafts will be published in Masanori Kaji, Helge Kragh, and Gabor Pallo, eds., *The Comparative Reception of the Periodic System* (Oxford: Oxford University Press, under contract).
16. See notes 11 and 12 above and *TDn*, vol. 2: 250.
17. "Honkai kiji," in *Tkk* 2 (1881): 14.
18. My translation. Sakurai 1882: 1.
19. My translation. Sakurai 1882: 1.
20. Sakurai 1883: 103f; Williamson 1851–1854.
21. On the *Dai-Nihon Kyōikukai* and other teachers' associations in prewar Japan, see Ueda 1954.
22. My translation. Sakurai 1884: 1.

23. See *Jpn. Calendar Tokyo Univ. Law, Sci., and Lit. 1881–1882*: 43–44. Sakurai listed Roscoe and Schorlemmer's *A Treatise of Chemistry* as one of the references.
24. Furukawa 1996: 92–94; Kikuchi 2000: 217f. For the European tradition of historical introductions, see, for example, Debus 1985: 1–2.
25. *Jpn. Calendar Tokyo Univ. Law, Sci., and Lit. 1883–1884*: 78.
26. "Philosophy of Chemistry" by Professor Joji Sakurai, Lecture Notes taken by Ichitaro Yokoji (A100.136), Tokyo Daigaku Sōgō Toshokan.
27. Kikuchi 2000: 218.
28. Kikuchi 2000: 223.
29. See note 20 above. Society proceedings show that Sakurai's presentation was originally given in English. See "Honkai kiji," *Tkk* 4 (1883): 7–10, on 7.
30. My translation. Sakurai 1885: 204. This article is partly included in Sugai and Tanaka 1970: 70–71. See also the record of his presidential address in "Honkai kiji," *Tkk* 6 (1885): 9–22, on p. 20.
31. My translation. Ibid.
32. Of 42 articles published in the *Tokyo kagaku kaishi* between 1880 and 1885, 35 were essentially analytical (including 28 manufacturing-oriented articles), and 7 were educational articles. See *Nihon kagaku kaishi hōbun sōmokuji narabini Sakuin* (1928).
33. Of 26 contributors to *Tokyo kagaku kaishi* between 1880 and 1885, 22 were graduates or former students of the Department of Chemistry in Tokyo. *Nihon kagaku kaishi hōbun sōmokuji narabini sakuin* (1928).
34. Kuhara 1882. Partly included in Sugai and Tanaka 1970: 71–72. On the date of his address, see "Honkai kiji," *Tkk* 3 (1882): 1–15, on pp. 4 and 7. Gotō 1980a: 28 summarizes this address.
35. Roscoe and Schorlemmer 1881–1892. Kuhara here referred to its part 1 published in 1881.
36. Kuhara 1882: 51.
37. Kuhara here referred to the original meaning of isomerism dating back to Berzelius (Berzelius 1830: 326f). This should not be confused with the current structure-based definition (geometrical isomers, stereoisomers, etc.) that would make Kuhara's discussion pointless.
38. My translation. Kuhara 1882: 54.
39. Kuhara 1882: 57–59.
40. Kuhara 1882: 59–60.
41. Kuhara 1882: 60–63.
42. My translation. Kuhara 1882: 62.
43. Takamatsu 1883a. The title of this address is my translation.
44. Takamatsu 1883a: 54.
45. My translation. Takamatsu 1883a: 48.

46. My translation. Takamatsu 1883a: 50.
47. Bud and Roberts 1984: 157; Donnelly 1987: 129–130.
48. See *Second Report of the Royal Commissioners on Technical Instruction* (1884), vol. 5: 106f.
49. "Honkai kiji," *Tkk* 4 (1883), pp. 6–7.
50. Tokyo Kōgyō Daigaku 1940: 57f.
51. Sakurai 1885: 204.
52. Shibusawa 1932.
53. For Shibusawa's business activities and discourse in the context of Japanese entrepreneurial history, see Hirschmeier 1964: 167–175 and Miyamoto 1999: 279–294.
54. On Shibusawa's activities at the Tokyo Fertilizer Company, see also Kamatani: 1989: 372–395.
55. Shibusawa 1932: 95.
56. *Takamatsu Toyokichi den*: 35 and 37.
57. *TDn*, vol. 2: 217.
58. *TDn*, vol 2: 359; *Takamatsu Toyokichi den*: 34–39. This is an example of the casual way in which important curricular changes were occasionally made at Tokyo University. As Terasaki Masao analyzed, decisions about the university's governance between 1881 and 1886 were officially made by a president, vice president and a secretary, plus the deans of the Faculties of Medicine, Law, Science, and Literature. Both the president and deans had consultative bodies, collectively called the Shijunkai, that included both senior and junior faculty as members, but decisions taken by these consultative bodies were not binding on the president and deans (Terasaki 2000a: 46–73). However, in this case Takamatsu skipped the Shijunkai and contacted Kato directly, leaving no record in the minutes of the Shijunkai: "Shijunkai kiji. Meiji 14 nen 9 gatsu," Tokyo Teikoku Daigaku Gojunenshi Bangō 45, Tokyo Daigaku Sōgō Toshokan.
59. *Jpn. Calendar Tokyo Univ. Law, Sci., and Lit. 1883–84*: 27.
60. My translation. *TDn*, vol 2: 364.
61. My translation. *TDn*, vol. 2: 359.
62. My translation. *TDn*, vol. 2: 221.
63. Takamatsu 1883b.
64. *TDn*, vol. 2: 335.
65. *Jpn. Calendar Tokyo Univ. Law, Sci., and Lit. 1883–84*: 26.
66. *TDn*, vol. 2: 439–475, on 462. Calculus, three-dimensional geometry, and differential equations were included in the mathematics course for second-year students in civil and mechanical engineering delivered by Miwa Kan-ichirō, which was extended to chemistry students in 1884. See also Miwa's report to the university for 1883–84 in *TDn*, vol. 2: 354.
67. My translation. *TDn*, vol. 2: 358–359.

68. My translation. *TDn*, vol. 2: 359. His student was Takashima Katsujirō, who graduated in 1884 from the Department of Pure Chemistry. Kikuchi 1999: 105.
69. Tanaka 1975: 63–64, Hirota 1988: 3–6, Bartholomew 1989: 79, Furukawa 1996: 93, and Kikuchi 2000: 225–226.
70. See *TDh, Tsūshi 1*: 500–504.
71. "Rigakubu chū bunkatsu shite sarani Kōgeigakubu o oki Hōgakubu o Hōseigakubu to Kaishō no ken," Sheet 27, *Meiji 18 nen kō* (A76), *Monbushō ōfuku*. "Tokyo Daigaku chū Kōgeigakubu o mōke sonota gakubu kaikō no ken" (2A-10-kō 3987: 64), *Kōbunroku*.
72. My translation, from ibid.
73. See, for example, Nakano 2003: 189.
74. Nakano 1995 and Nakano 2003: 161–162.
75. Nakano 2003: 161–162.
76. This source was transcribed in Ōkubo 1972, vol. 1: 351–356. For the dating of *Gakusei yōryō*, see Hall 1973: 410–412 and Inuzuka 1986: 265–266. This document was first mentioned in Ōkubo 1944: 111–120.
77. My translation, from Ōkubo 1972, vol. 1: 356.
78. Ōkubo 1972, vol. 1: 351 and 353.
79. Hall 1973: 411.
80. See Nakano 2003: 195.
81. See, for example, Swale 2000 in addition to Hall 1973 and Inuzuka 1986.
82. See Ōkubo 1988: 135–139.
83. Ōkubo 1976: 15. Nishi's "Chisetsu" where his coinage of *kagaku* first appeared in print was published in 1874 in *Meiroku zasshi*, the organ of the *Meirokusha*.
84. My translation, from "Chisetsu: yon," in Yamamuro and Nakanome 1999–2009, vol. 2: 233–238, on 236f; Tsuji 1973: 181. See also chapter 1 above, note 62.
85. My translation, from "Kōbu Shōyū jidai Ikensho Fukumeisho," *Watanabe Hiromoto Monjo* 53, Tokyo Daigakushi Shiryōshitsu. Cited in Nakano 2003: 61.
86. See, for example, Kakihara 2002: 75. It discusses the reduction of practical training in factories in large parts of the Imperial College of Engineering later in its existence. This attests to a discrepancy between Watanabe's rhetoric and institutional reality in the 1880s.
87. There is an extensive literature on the early years of the Imperial University as defined in the Imperial University Ordinance on the initiative of Mori. See Nakano 2003: 78f and 194f.
88. Terasaki 2000a: chapter 3 (pp. 253–340), Nakayama 1978: 130–169, and Bartholomew 1989: 104f. and 134.
89. Watanabe 1886.

90. There is no extensive biography of Watanabe. See, however, his curriculum vitae in Tokyo Daigakushi Shiryōshitsu 2005: 4.
91. *TDn*, vol. 3: 134.
92. See Watanabe's letter to Sakurai dated October 16, 1886, and included in the latter's bequest in 2002 ex Yamamoto Kazuko (*Jirei na*), Sakurai Jōji Papers. It is probably not a coincidence that Watanabe asked three British-trained chemists to take on this task, because the gaslight industry was first developed in Britain in the 1820s and then spread to the Continent (Tomory 2011).
93. Bartholomew 1989: 133 made this point.
94. Nakano 1999: 132–138.
95. Amano 2005: 50f.
96. *TDn*, vol. 5: 76 and 297f; vol. 6: 210f., 376f., and 509f. See Miyamoto 1999: 137–235 for the role of *seishō* in economic and entrepreneurial history in Meiji Japan.
97. *TDn*, vol. 3: 36.
98. *TDn*, vol. 5: 116, 141, 453, and 493; *TDn*, vol. 6: 61, 152, 316, 364f, and 553.
99. *TDn*, vol. 3: 233. Each year two positions each were provided for students of literature and science. They included both tuition waiver and maintenance grant. As I discuss in chapter 6, the simplified course of the College of Science was established in 1889 by the same demand by the Ministry of Education.
100. Terasaki 2000a: 158–161.
101. Nakano 2003: 162–165.
102. Notification from Tokyo Imperial University to Sakurai dated June 2, 1911, and included in the latter's bequest in 2002 ex Yamamoto Kazuko (*Jirei ke*), Sakurai Jōji Papers.
103. Sakurai 1940: 18–19.

5 Constructing a Pedagogical Space for Pure Chemistry

1. It was renamed Tokyo Teikoku Daigaku (Tokyo Imperial University) in 1897 when the second imperial University, Kyoto Imperial University, was established. Tokyo Imperial University retained this name until 1947, when it returned to an early Meiji name, Tokyo Daigaku, after the collapse of the Empire of Japan.
2. *TDn*, vol. 5: 137.
3. Tanaka 1975: 37.
4. Sakurai's colleagues and acquaintances "considered him [Sakurai] the essential English gentleman because he was dignified, well-mannered, and always impeccably dressed." Bartholomew 1989: 180. Bartholomew used other attributes such as "elegant and imposing" and "strict and slightly aloof" for Sakurai.

5. See Michel Foucault's analysis of the panopticon, in Foucault 1979: 200–203. See also Markus 1993: 39–168. On the relevance of Foucault's approach to the studies of spaces for science, see Smith and Agar 1998: 1–23.
6. Bracken 2013: 110–113 and 162–165.
7. Brock 1989: 164.
8. Gooday 1991: 99–104.
9. Johnson 1989: 223.
10. Bartholomew 1989: 96.
11. Nakano 2003: 14–19.
12. *TTDg*, vol. 2, has an appended the campus map of the Faculty of Medicine in the Hongō campus for 1880 and the Faculties of Law, Science, and Literature on the Kanda-Nishikichō campus for 1878. There is no independent laboratory building on Hongō campus maps of the time, which means that the Faculty of Medicine should have used other rooms in its buildings as laboratories. The three separate chemical laboratories on the Kanda-Nishikichō campus were a quantitative analysis laboratory, qualitative analysis laboratory (built in 1879), and applied chemistry laboratory (built in 1882 or 1883). According to the plan of the main Kanda-Nishikichō building (illustrated in figure 2.1) in 1875, the building had one room for engineering (*kōgakushitsu*) and another for science departments (*rigakushitsu*). *Kaisei gakkō kenchiku zumen zenmen* (no. 36). *Kōbun fuzoku no zu,* Kokuritsu kōbunshokan.
13. Kanao 1960 remains a standard biography of Nagai, and its usefulness is augmented by Nagai's letters and autobiography reproduced there. Care should be taken, however, in using this source because of its hagiographic nature as the author was a former student of Nagai.
14. Nakano 2003: 7–10.
15. Kanao 1960: 138. See also Nishizawa 1923: 255, on the collaboration of Yamaguchi and Nagai in the designing of other buildings of the Faculty of Science of Tokyo University.
16. Jackson 2008a: 225–275 is an excellent account of the technical development of chemical laboratories in the nineteenth century as an essential tool for synthetic organic chemists. Jackson identifies the Bonn and Berlin laboratories, both designed by Hofmann, as the most famous examples.
17. *Yamaguchi Hakushi Kenchiku Zushū* (n.d.). Hofmann 1866: 56–57, 60–61, and 64–65.
18. Jackson 2008a: 262.
19. It has not been addressed by historians why German (or more broadly Germanic, including, for example, Austrian, Bohemian, and Swedish) chemical institutes generally included a director's residences whereas their counterparts in the rest of Europe and North America did not, as observed by Peter Morris. See "The Director's Residence

and the Spy Window," in Morris, *Crucible of Science: The History of the Chemical Laboratory* (London: Reaktion Press, forthcoming). Whatever its origins, this aspect of laboratory design undoubtedly came partly from all-powerful and life-long directorship in Germanic chemical institutes. I am grateful to Peter Morris for sharing his manuscript.

20. For the urban designs of Tokyo in the Meiji period, see Fujimori 2004. Yamaguchi was a pioneering professional urban designer in Meiji Japan with broad training in civil engineering, architecture, public administration, and political economy at the Ecole Centrale des Arts et Manufactures in Paris (Fujimori 2004: 330–331).
21. Fujimori 2004: 279–281. David Wittner interpreted the importation of Western material culture, including architecture, into early Meiji Japan as part of an attempt by "cultural materialists" to form an ideology of "progress" (Wittner 2008: 99–124).
22. See, for example, Brock 1997: 48–51.
23. Hentschel 2002: 47–55.
24. See chapter 3 above, note 107.
25. Johnson 1989: 223.
26. *TDh, Bukyokushi 2*: 480f.
27. *TDn*, vol. 3: 133. See also "Rika Daigaku Kyōshitsu Rakusei su," *Kanpō*, no. 1649 (December 26, 1888): 278. University President Watanabe Hiromoto confirmed in 1888 that the reorganization of the university had been the reason for changing the plan of a chemical laboratory into that of the Main Science Building. See the minutes of the Imperial University Council of May 21 and 23, 1888, in *TDn*, vol. 3: 126.
28. Nakano 2003: 63. Nagai was among the "off-duties" because he was out of the country to marry a German fiancée (Kanao 1960: 89–94). There have been considerable debates about the "true" cause of Nagai's dismissal beyond what Nakano clarified. James Bartholomew asserted that "in November 1885 [Sakurai] secured Nagai's dismissal from the university on the grounds that his [Nagai's] interest in pharmacological chemistry was not academic in nature" (Bartholomew 1989: 79). Kanao Seizō understandably tried to explain this dismissal in favor of Nagai and resorted to a kind of conspiracy theory by pointing a finger at the "British faction" of Sakurai and Kikuchi Dairoku, Cambridge-trained mathematician and dean of the Faculty of Science at the time (Kanao 1960: 138). Similarly, Kaji Masanori cited the rumor of a "Sakurai-Nagai quarrel" among students, according to which Sakurai deleted Nagai's name from the list of prospective professors for an Imperial University in March 1886 due to Sakurai's strong Germanophobia (Kaji 2011a: 182, n. 21). Yasue Masakazu, in contrast, denied any involvement of Sakurai in Nagai's dismissal (Yasue 1983 and 1986). A major obstacle to any judgment

on this debate is that there is no other source except rumors, such as the ones cited by Kaji, to underpin either argument. The following discussion does not participate in this debate but concentrates on the change in laboratory design made by Sakurai and Divers.

29. *TDn*, vol. 5: 462. The fact shown here that *both* Divers *and* Sakurai were involved with designing the interior of the building is important in considering the question about the comparative status of Japanese and non-Japanese faculty at Tokyo. In fact, different Japanese words were used to designate the job titles of Japanese and non-Japanese professors at Tokyo University and the Imperial University from around 1880: i.e., *kyōju* and *kyōshi*, respectively. Before 1880, all (mostly foreign) professors were *kyōju* (compare *TDn*, vol. 1: v–vi; and vol. 2: iv–v). This timing corresponds to the period when foreign professors started to be replaced by Japanese professors at Tokyo University (see chapter 4). Does that mean the Japanese and non-Japanese faculty were treated differently by Tokyo University, the Imperial University, and its constituent colleges and departments from then on? On the one hand, it strongly suggests the diminished role of foreigners in university and college management, as shown by the two facts that Divers did not officially occupy a college chair and that he did not attend the Senate of the College of Science, both instituted in 1893 (see, e.g., *Jpn. Calendar (Tokyo) Imp. Univ. 1893–94*: 199 and the minutes of the Senate of the College of Science at (Tokyo) Imperial University, Tokyo Daigaku Rigakubu Jimubu Shomu kakari). Also, it is likely that Sakurai held responsibility over department finance (see chapter 4, note 92). On the other hand, in addition to designing the interior of the department building, Divers was asked by President Watanabe to join discussion of the installation of facilities for manufacturing coal gas and a steam engine for generating electricity, both of which suggest that Divers's influence on department management and, more important, on access to spaces and other teaching and research resources was at least comparable to that of his Japanese colleagues. Moreover, in faculty rosters for the College of Science included in university calendars, Divers was always given the first place next to the college dean and vice dean, no small feat in a society where orders or ranks were so important (see, e.g., *Jpn. Calendar (Tokyo) Imp. Univ. 1886-87*: 94). In short, an answer to the question of status seems to depend on the level of the university hierarchy (university, college, or department) and on the specific activities (administration, management, or academic teaching) we are talking about. I am grateful to Masao Uchida for drawing my attention to this question.
30. Chapter 4, notes 91 and 92 (coal gas) and Majima 1954: 4 (equipment).
31. Iimori Satoyasu, who drew the sketch on which figure 5.6 is based, graduated from the department in 1910. That means Iimori was a

student after Divers retired in 1899 and before the relocation of the Department of Chemistry into another building in 1916. In interpreting this source, I take these circumstances into account and have inferred that Haga Tamemasa, the successor of Divers, took over Divers's office and private laboratory in 1899, and that Ikeda, who succeeded Haga as assistant professor in 1896, occupied Haga's place, both without changing the fundamental spatial structure.

32. As was shown by the illustration by Wilhelm Trautschold, in Brock 1997: 54f.
33. See a drawing of the Birkbeck Laboratory in *The Illustrated London News*, May 30, 1846, where an unidentified artist emphasized natural light, which is most beneficial in this alignment of a working bench. On the RCC lab, see Jackson 2008a: 236.
34. The wording of "senior" and "junior" professors here does not necessarily correspond to their titles (*kyōju/jo-kyōju* [full/assistant professors]), but to the original meaning of who was appointed earlier.
35. My translation. *TDn*, vol. 6: 327.
36. *TDn*, vol. 6: 328.
37. Kikuchi 2012: 295, 310 (n. 49), and 311 (n. 50).
38. Chapter 3, note 69.
39. See the plan of the Birkbeck Laboratory in *An Account of the New North Wing and Recent Additions to University College, London,* [...] (1881).
40. Chapter 3, note 69.
41. Majima 1954: 3–4.
42. See his mentions of Haga in Ōsaka 1949: 490 and 492 and Majima 1954: 4–5.
43. For Haga's life, see Shibata 1961: 784ff. In contrasting Haga's character with Sakurai's, Bartholomew used such adjectives as "rumpled and ordinary," "open," "readily accessible" and "erudite but guileless," but failed to mention the cultural aspect of Haga's "accessibility" and "erudition" (Bartholomew 1989: 181). According to my interpretation, the contrast between Sakurai and Haga is more appropriately described with ***haikara*** (refined in the Western way) and ***bankara*** (rough and unrefined in the Japanese way), which I used in chapter 2.
44. Majima 1954: 4. The importance of glassblowing in nineteenth-century chemical pedagogy is discussed in Jackson 2008a: 162–224.
45. Tanaka 1975: 37f. Partly cited in Bartholomew 1989: 181 (n. 81).
46. See Servos 1990: 95–99 and Kikuchi 2000: 231.
47. Ostwald 1897 and 1899.
48. Haga and Ōsaka 1895; Haga and Majima 1903.
49. Ikeda and Kametaka 1899. See also Katayama 1907, in which Katayama acknowledged Ikeda's help, but not Sakurai's.
50. See Majima 1954: 6.

51. Later in his life, Sakurai gave an address as acting president of Tokyo Imperial University at its entrance ceremony in October 1912: "You should not only attend lectures and digest and assimilate them, but also visit the library or research rooms frequently and consult reference tools and scholarly journals, both domestic and international, to acquire knowledge from all over the world. You should also attend lectures or read books with an investigative attitude (*kenkyū tekino taido* [sic]). If you have any questions, you should ask your teachers and have the courage to debate with them if necessary." (My translation, in Sakurai 1940: 89–92, on 90). Though not contemporary direct evidence of his motives in setting up the *Zasshi-kai*, the quotation is an interesting retrospective reflection on Sakurai's view of what should have been the student's experience.
52. Samejima 1942: 126 and *TDh, Bukyokushi 2*: 441.
53. Bartholomew 1974 and Bartholomew 1989: 190.
54. James Bartholomew located the origin of the *Zasshi-kai* in the section of the Department of Chemistry at Tōhoku Imperial University in Sendai established in 1911 (Bartholomew 1989: 187). However, the chronology and common Tokyo academic background of the three founding chemistry professors at Sendai (Masataka Ogawa, Majima, and Masao Katayama) make it clear that the *Zasshi-kai* had its origin in Tokyo's Department of Chemistry. Majima and Katayama mentioned Tokyo's *Zasshi-kai* in their recollections (Majima 1954: 4 and Katayama 1950: 63). Majima explicitly stated that the *Zasshi-kai* was a special characteristic of the Department of Chemistry at Tokyo.
55. The first page of an untitled notebook, "Notebook 2," in Sakurai Jōji Papers: C. See chapter 6, table 6.1 for an annotated catalogue of Sakurai's notebooks.
56. *Kagakuka Zasshi-kai kiji*, the Japanese-language manuscript proceedings of Tokyo's *Zasshi-kai*. 25 vols. Tokyo Daigaku Rigakubu Kagakuka.
57. "The Journal Meeting Minutes" in Sakurai's untitled notebook 2 (Sakurai Jōji Papers: C) recorded only the first meeting held on January 28, 1890, and no other meetings are mentioned in this source thereafter. The first recorded proceedings of the Chemistry Alumni Society on December 2, 1892, (*Kagakuka Zasshi-kai kiji*, vol. 1) mentioned that this organization had been on the verge of disappearance since its establishment in 1890 and added that the entrance of Sakurai had altered this stagnant mood.
58. Proceedings for January 5 and February 4, 1893, *Kagakuka Zasshi-kai kiji*, vol. 1.
59. The two-volume proceedings of the Departmental Conference, which was included in the proceedings of *Zasshi-kai* as volumes 17 and 18, shows that the financing of this conference was treated as a part of the *Zasshi-kai*, which means that the Departmental Conference was

organized by *Zasshi-kai*. On this conference, see also Sasaki and Tachibana 1991, a list of graduate theses presented at this conference.

60. Bartholomew 1989: 180.
61. The leisure culture of imperial university students originated in *konpa*, derived from the English word "company," in the late 1880s and 1890s in the First Higher School (*Dai-ichi Kōtō Gakkō*), the representative preparatory state boarding school for those aiming at admission to imperial universities. *Konpa* means room parties, largely in the rough *bankara* style; they were frequently organized by students in higher-school dormitories to strengthen group solidarity and to impose it on freshmen. See Roden 1980: 110f.
62. Bartholomew 1989: 181. Readers are reminded that Bartholomew's negative appraisal of Haga is based on this phase of his career as a lecturer and overlooks Haga's earlier productive role as an experimental trainer.
63. The Kanto Earthquake in 1923 destroyed many other contemporary buildings on the Hongō campus, including the Main Science Building. However, the new Chemistry Building survived, and with extensions and renovations, it still serves Tokyo's Department of Chemistry. See Kinoshita, Ōba, and Kishida 2005.
64. *Ayumi*: 92. See also Samejima 1956: 109.
65. *Ayumi*: 92–95. The Main Science Building had ca. 1,822 square meters of floor area, but the Department of Chemistry occupied only its second floor, whose available area was further diminished by courtyards. The new Chemical Institute had ca. 1,260 square meters with two floors and a basement, all of which were available to the department. For details and building data of the two laboratories, see *TTDg*, vol. 2: 1270 and *TDh, Bukyokushi 2*: 481–484. In *TTDg*, the Main Science Building is designated as "The Department of Physics Building" because the Department of Physics took over the whole building after the relocation of the Department of Chemistry in 1916.
66. Ostwald 1898: two folded plans *Tafel I* and *Tafel II*.
67. One of such rare occasions was the visit in June 1936 of Francis William Aston, the British chemist and physicist who won the 1922 Nobel Prize in chemistry for his discovery of isotopes and the whole-number rule. The picture taken on this occasion depicts Aston among senior professors of both chemistry and physics at Tokyo in front of the open façade of the Chemical Institute (*Ayumi*: 57).

6 Making Use of a Pedagogical Space for Pure Chemistry

1. Letter from Sakurai to Wilhelm Ostwald dated June 14, 1898, a part of *Nachlass Ostwald* in possession of the Akademiearchiv, Berlin-Brandenburgische Akademie der Wissenschaften, Germany. This letter is transcribed in Kikuchi 2000: 248–252.

2. Previous discussions of the Department of Chemistry at Tokyo include Bartholomew 1989: 179–182, and Tanaka 1975. For official departmental histories, see Samejima 1942; "Kagakuka," in *TDh, Bukyokushi 2*: 432–484; and *Ayumi*. Publications of chemistry-related societies such as Nihon Bunseki Kagakukai Sōritsu Jusshūnen Kinenshi Henshū Iinkai 1963, Nihon Kagakukai 1978, and Nihon Bunseki Kagakukai 1981 also mention the teaching at Tokyo's Department of Chemistry.
3. *TDn*, vol. 6: 329.
4. Sugawara 1978: 189f.
5. My translation, from Tanaka 1975: 40.
6. Bartholomew 1989: 180.
7. *TDn*, vol. 6: 164. He made similar statements about his lectures on chemical philosophy and, later, on theoretical and physical chemistry. Kikuchi 2000: 217.
8. Sakurai Jōji Papers: C (lecture notes).
9. This source is hereinafter referred to as "Sketches."
10. See for example, Lecture 76 (Constitution of isatin: criticism of the hydroxyl formula) and Lecture 89 (Hantzsch's view of the constitution of pyridin, its criticism) in the earliest syllabus for organic chemistry in "Sketches."
11. See *Jpn Calendar Tokyo Univ. Law, Sci., and Lit. 1881–1882*: 43–44.
12. See, for example, Lectures 21, 33, 35, and 80 in Sakurai's earliest syllabus for theoretical and physical chemistry in "Sketches." The collation between this syllabus and Untitled Notebook 1 shows that Sakurai delivered Lectures 19–39 under the heading "Relation Between Atomic Constitution and Physical Properties."
13. Lectures 53–66, in Sakurai's earliest syllabus for theoretical and physical chemistry in "Sketches."
14. Kikuchi 2000: 251f.
15. Sakurai 1888: 439–440 and Sakurai 1899: 8.
16. For Sakurai's impact on Ikeda, Ōsaka, and Katayama, see, for example, Kikuchi 2000 and Kikuchi 2005. See also Ōsaka 1949: 111 and Katayama 1950: 64.
17. On Majima's research, see Kaji 2011a and Kaji 2011b.
18. Tanaka 1975: 41.
19. My translation. Tanaka 1975: 39. Though Sakurai was by then no longer lecturing routinely, he substituted for Matsubara Kōichi who was absent on study leave in Manchester and Berlin while Shibata was a student.
20. See Lecture 72 of the earliest syllabus of Sakurai's lectures on organic chemistry, in "Sketches."
21. Tanaka 1975: 56–61 (Shibata's early research in organic chemistry) and 83–130 (his change of specialism to complex chemistry). Sakurai's recommendation was made on the basis of job opportunities as the chair of inorganic chemistry at Tokyo fell vacant in 1910.

22. As related by himself at the Departmental Conference for Presentation of Graduate Theses on February 24, 1934. See the second volume of the proceedings of the conference, classified as volume 18 of the Proceeding of the *Zasshi-kai*. For reference see chapter 5 above, note 56.
23. Nakano 1999: 117.
24. My translation. Ōsaka 1949: 489f.
25. My translation. Majima 1954: 3.
26. My translation. *TDn*, vol. 6: 173.
27. See, for example, *ICET Calendar 1873–1874*: 16f and *ICET Calendar 1875–1876*: 35f.
28. This anecdote was related by Ikeda himself at the Departmental Conference for Presentation of Graduate Theses on February 24, 1934. See the second volume of the proceedings of the conference, classified as volume 18 of the proceeding of the *Zasshi-kai* (see chapter 5 above, note 59).
29. *Divers Kyōju, Ikeda Kikunae Kyōju no Shiken Mondai, Tōan*, Tokyo Daigaku Rigakubu Kagakuka. It is in contrast with the examination papers of Ikeda, who was appointed full professor in 1901, in a later period in this collection, which were all written in Japanese. This collection is briefly mentioned in *Kagakushi* 20 (1993): 207. To look at Divers's lectures on inorganic chemistry, the examination papers of S. Hada, T. Ichioka, and Y. Horiike, all classified in the year 1890, proved particularly useful.
30. Haga 1912: 11f.
31. For nineteenth-century organic structural theory and stereochemistry, see, e.g., Ramberg 2003 and Rocke 2010.
32. Divers was granted the title of "emeritus teacher" (*meiyo kyōshi*) upon retirement; he returned to England in 1899. One year after his departure a statue of Divers was erected in front of the main building of the College of Science, a donation of his former students and colleagues. Upon his death in 1912, a memorial service for Divers at Tokyo Imperial University attracted more than 60 mourners consisting of former students and colleagues (see *Tokyo kagaku kaishi* 33 (1912): 539–541).
33. Ōsaka 1949: 4.
34. Haga 1912: 10.
35. Ōsaka 1949: 4.
36. Kikuchi 2012: 297–302.
37. For this analysis, I used the list of papers from the Department of Chemistry for the period between 1877 and 1940 (pp. 137–187), appended to Samejima 1942, and the list of Divers's papers in Haga 1912: 18–26, together with Royal Society of London 1867–1925. The papers related to Divers's supervision of students' works at the Imperial University amount to 13 in total.

38. Divers and Haga 1896.
39. Divers and Hada 1899, Divers and Ogawa 1899, 1900a, 1900b, 1901, and 1902.
40. Hada 1895, 1896, and 1897; Chikashige 1895 and 1897; and Haga and Ōsaka 1895.
41. Divers and Haga 1896: 1653. "We are indebted, for their kind assistance, to Mr. Y. Osaka, B.Sc., in examining the reactions of silver amidosulphonate, and to Mr M. Chikashigé, BSc, in examining the compound of amidosulphonic acid with sodium sulphate."
42. This summary was prepared because the work was submitted to the College of Science for examination for the award of a Doctor of Science degree (*Rigaku Hakushi*) to Haga in 1894, in which Sakurai was the examiner. Statement of Divers addressed to Sakurai, dated February 19, 1894. Appended to the minutes of the Senate of the College of Science at (Tokyo) Imperial University, February 23, 1894. *Tokyo Teikoku Daigaku Rigakubu Kyōjukai Kiroku: Meiji 26 nen 9 gatu yori Meiji 45 nen 6 gatu ni itaru,* Tokyo Daigaku Rigakubu Jimubu Shomu kakari.
43. Haga 1912: 8.
44. See Divers and Haga 1896, Sakurai 1896, and Loew 1896. Takahashi's contribution was noted in Divers' addenda (appended to Loew's paper) in pp. 1664–1665.
45. Tanaka 1975: 41 and Bartholomew 1989: 181.
46. Kikuchi 2012: 299.
47. See Yoshihara 2000 and Yoshihara 2008 for Ogawa's work on Nipponium.
48. See Ogawa's recollection of his joint research with Divers, in Ono and Fujise 1961: 222. The word "orthodoxy" (*seitō*) is Ogawa's own as quoted by one of his students, Ono Heihachirō.
49. Yoshihara 2008: 233–235.
50. On Ogawa and his collaborators, see Yoshihara 2003.
51. Ikeda 1890: 268.
52. See Lectures 28 and 29 in the first syllabus of Sakurai's course of theoretical and physical chemistry in "Sketches." That Sakurai inserted "Ikeda's expression" into the original syllabus, presumably in 1895, is clear from the shade of the ink Sakurai used.
53. It also seems likely in the following case that students took their research topics from Sakurai's lectures: Ikeda 1898a, Ikeda 1898b, Ikeda 1899, Ikeda and Kametaka 1899, and Matsubara 1899. The following are the research topics of these papers and the places where they were mentioned in Sakurai's lectures seen in the first syllabuses in "Sketches": the nature of osmotic pressure (Lecture 46, theoretical/physical chemistry); reaction rate of the hydrolysis of esters (Lecture 75, theoretical/physical chemistry); boiling points of homologous series (Lecture 20, theoretical/physical chemistry); measurement of

molecular points using the elevation of boiling points (Lecture 48, theoretical/physical chemistry and Lecture 1, organic chemistry).

54. That Sakurai helped his former student and assistant professor, Yoshida Hikorokurō, in preparing English papers is mentioned in Yoshida 1883: 473 and Yoshida 1885: 801. It would have been surprising if Sakurai had not performed the same service for his students at the Imperial University.
55. Ikeda 1898a: 142.
56. Sakurai mentioned this in his letter to Wilhelm Ostwald dated June 14, 1898 (note 1 above). Transcribed in Kikuchi 2000: 251f.
57. Samejima 1942: 127, Sakanoue 1997: 163, and Bartholomew 1989: 180.
58. Sakurai 1892a.
59. Sakurai 1892a: 508.
60. Sakurai: 1892b. For Beckmann's method, see, for example, Beckmann 1889.
61. Sakurai 1892b: 995.
62. Liebig's *kaliapparat* is one of the most vigorously discussed apparatuses in the history of chemistry. See, for example, Holmes 1989 and Usselman et. al. 2005.
63. For a concise historical account of molecular weight measurement using colligative properties of solutions, see Campbell 1998, which does not mention Sakurai.
64. Sakurai 1894 and Sakurai 1895a. The full paper is Sakurai 1895b.
65. Sakurai 1896: 1660f.
66. Sakurai 1892a: 508.
67. Matsubara 1899: 600–602.
68. Jackson 2008b.
69. My translation. Shibata 1939: 113.
70. Tanaka 1975: 91.
71. "Organische Chemie von J. Sakurai, Zweiter Band: Verbindungen der Aromatische Reihe," (4/M394) in Majima Bunko, Osaka Daigaku Fuzoku Toshokan.
72. Majima 1954: 3–4 and Kaji 2011a: 175–176.
73. Kikuchi 2005.
74. Ikeda 1909. Partly reproduced in Sugai and Tanaka 1970: 161–163. Ikeda 2002 is an abridged English translation. See also Hirota 1994: 139–171. Morris-Suzuki 1994: 118–119 devotes two paragraphs to Ikeda's research as an example of laboratory-based original invention. Sand 2005 discusses the sociocultural and economic context of Ajinomoto's success and later problems.
75. Ikeda 1956: 223. Hirota questioned the validity of the source regarding how exactly Ikeda became interested in umami as a research topic, but he does not question this part on Ikeda's general interest in chemical technology (Hirota 1994: 197–200).

76. Ikeda 1902. Partly reproduced in Sugai and Tanaka 1970: 73–75.
77. Ikeda 1902: 1145.
78. Ikeda 1902: 1147 and 1150–54. See also Servos 1990: 63–65 (physical chemistry as *allgemeine Chemie*) and 66–70 (physical chemistry as a "donor specialty"). John Servos pointed out that the first characterization of physical chemistry as *allgemeine Chemie* remained propaganda rather than a reflection of reality in the early years of physical chemistry's development at the turn of the century. Most physical chemists were involved in detailed puzzle-solving left by Van 't Hoff, Arrhenius, and Ostwald.
79. Hirota 1994: 158–160. As Hirota pointed out, it is more precisely the *levo*-form of sodium glutamate. See also Brock 2011: 13–15. Ikeda mentioned the "flavorless" taste of glutamic acid described by German organic chemist Emil Fischer and added that its sodium salt was ignored by him (Ikeda 1909: 827–828; Ikeda 2002: 848r). The failure to try monosodium glutamate as a likely source of umami by agricultural chemist Suzuki Umetarō, a student of Fischer, was recounted in Katayama 1956: 38.
80. Ikeda, 1909: 829–830 (Ikeda 2002: 849l).
81. Ikeda 1909: 835–836 (Ikeda 2002: p. 849r). Hirota 1994: 160–169. For the date of his patents, see *Ikeda Kikunae tsuiokuroku* (1956): 219–222, on p. 219.
82. Hirota 1994: 37–38.
83. Ikeda 1915: 19–20 and note 45 above. Compare his entirely positive assessment of Divers in Ikeda 1904: 480–482.
84. Ikeda 1956: 224.
85. Kikuchi 2000: 229–236. For a list of the first occupation of the graduates from the Department of Chemistry between 1867 and 1926, see Kikuchi 1999: 105–111.
86. My translation. *TDn*, vol. 3:91.
87. Watanabe 1886.
88. *TDh, Tsūshi 2*: 138f.
89. Ōsaka 1949: 494.
90. Ōya and Hara 1965: 192 (his role in the screening of textbooks) and Kikuchi 2000: 234 and 242, n. 69 (on curriculum design).
91. Kikuchi 2000: 230 (n. 62 and 63). For Frankland's course, see Brock 1993: 407 and Russell 1996: 294–298.
92. Majima 1954: 146. See also Ōsaka 1949: 492–494 and 550, Majima 1954: 6–7 and 10 (which also mentioned the roles of Sakurai and Majima in Katayama's appointment at Sendai) and Tanaka 1975: 65 and 75 (on Shibata).
93. Hirota 1994: 36.
94. See, for example, Tsuda 1956.
95. See note 75 above.

7 Connecting Applied Chemistry Teaching to Manufacturing

1. My translation. Takamatsu 1895a: 189.
2. Nakamura was replaced by Shizuki Iwaichirō, another student of Divers in March 1887.
3. See, for example, Kakihara 2002: 75–76.
4. My translation, from Nakazawa 1932: 37, reproduced in Tanaka and Yamazaki 1964: 79–80.
5. Matsui was appointed professor and vice-principal in 1887 at the Third Higher School in Kyoto, and he returned to the Imperial University in 1890 as an imperial university councilor and dean and professor of chemistry of the newly created College of Agriculture, which arguably offered him a better opportunity to use the expertise in physiological chemistry and biochemistry he had developed at Columbia, Yale, and Tokyo University. One of his students at the College of Agriculture, Kozai Yoshinao, recalled Matsui's teaching prowess shown in his detailed yet clear-cut lectures in chemical theories but pointed out that he was always too busy with administrative matters to do research until his untimely death in 1911. Kozai 1911: 2–3 and 5.
6. Nakaoka 2006: 169–171.
7. The chemico-technological education of Takamatsu and Nakazawa at the Imperial University is a neglected topic in the history of science, technology, and education in Japan. See, however, Tanaka and Yamazaki 1964: 78–80, which briefly explains the establishment of the Department of Applied Chemistry; see also Kamatani 1989 for the history of chemical industry in Japan, especially inorganic heavy chemicals and Nakazawa's consulting work; see Takamatsu 2001 for a historical survey of Japanese chemical industry. See also Nakaoka Tetsuō's perceptive comment that Takamatsu and other former overseas students played an important role in "helping transferred technology take root in Japanese geographical and cultural conditions and encouraged innovations by Japanese" (Nakaoka 2001: 23). For technological education, see Miyoshi 1979 and Miyachi 2001, although they do not cover chemical education.
8. Based on "Nakazawa Iwata Hakushi den (Biography of Dr. Nakazawa Iwata)" in *Nakazawa Iwata kinenchō*: 61–102 unless indicated otherwise.
9. Kumazawa 2004.
10. For biographical data on Wagener, see Ueda 1925, Umeda 1938, Yorita 1984, Shiba 1999, and Ozawa 2011.
11. Shiba 2006: 59–64.
12. *TDn*, vol. 2: 155.
13. Ibid.

14. *TDn*, vol. 2: 249f.
15. Nakazawa 1938: 87.
16. *Nakazawa Iwata kinenchō*: 69.
17. Sheets 153 and 217–221, *Ryugakusei Kankei Shorui: Meiji 17, 18 nen* (G13); sheets 30–37, *Ryugakusei Kanren Shorui: Meiji 19 nen yori Meiji 24 nen ni itaru* (G14). Tokyo Daigakushi Shiryōshitsu.
18. On the Krefeld school, see *Second Report of the Royal Commissioners on Technical Instruction* (1884), vol. I: 133–141. See also Lundgreen 1987 and Lundgreen 1990: 41–46 for German *Fachschulen* and more widely defined German engineering education. The character of the Weihenstephan agricultural academy with a strong emphasis on practical work and experimentation on farms, in comparison with the more academic Faculty of Agriculture of the University of Munich, is thoroughly discussed in Harwood 2005: 175–221.
19. Sugawara 1978: 189f. According to Nakazawa's report in Japanese, this course dealt with "analytical techniques of industrial importance that are not practiced in the ordinary course of analytical chemistry," including spectroscopic qualitative analysis and gas analysis. See *TDn*, vol. 6: 231, 339 and 535.
20. *TDn*, vol. 6: 225.
21. Inoue 1932: 50.
22. Mizuta 1932: 426.
23. *Nakazawa Iwata kinenchō*: 53.
24. Nakazawa 1882, 1883, 1888, 1889, and 1892.
25. *TDn*, vol. 6: 231.
26. *TDn*, vol. 6: 231f and 399.
27. *TDn*, vol. 6: 399.
28. Takamatsu 1895a.
29. *TDn*, vol. 5: 100 and 315; *TDn*, vol. 6: 231f., 399, and 534f.
30. *Nihon Kagaku Kaishi Hōbun Sōmokuji narabini Sakuin* (1928).
31. *Teikoku Daigaku Gakujutsu Taikan: Kōgakubu, Kōkū Kenkyūjo* (1944): 23f.
32. See, for example, Tanaka 1961a: 707 and Mizuta 1932: 427–428.
33. See the recollection of Nishikawa Torakichi and Shinozaki Yūzō, in *Nakazawa Iwata kinenchō*: 38–42 and 52–54, on 41 and 53.
34. *Jpn. Calendar Imp. Univ. 1886–1887*: 77.
35. My translation, from *TDn*, vol. 6: 232.
36. My translation, from *TDn*, vol. 6: 399.
37. Using college annual reports, it is possible to document this until 1889. The College of Science dispatched Yoshida to several provincial areas in July 1887 and in July of the following year to collect raw material of *urushi* (Japan) for his chemical study. Divers and Haga also travelled with the same scheme in March 1888. It is indeed

possible to infer that this practice continued after 1889. *TDn*, vol. 5: 454; and vol. 6: 153f.

38. *TDn*, vol. 5: 77–75 and 298–299; *TDn*, vol. 6: 211, 376–377, and 510–511.
39. This is in stark contrast with the career patterns of the graduates of the Department of Chemistry at the College of Science during the Meiji period, which were dominated by teachers. See Kikuchi 2000: 231–236.
40. See Tokyo Kōgyō Daigaku 1940: 1117 and 1132. It mentions Nakano Hatsune and Yamagawa Gitarō in electric engineering and Taniguchi Naosada in mechanical engineering.
41. Takamatsu 1891, Takamatsu 1892, Takamatsu 1893, and Takamatsu 1895b. *Ranjō seizō shiken hōkoku* (1895) mentioned that his investigation of indigo was commissioned by the Ministry of Agriculture and Commerce. The last work was a sequel to Takamatsu's papers, executed by an analyst of the ministry. Ikeda Kikunae rated this research highly and asserted Takamatsu's role in the dyeing industry in Japan (Ikeda 1904: 482–483).
42. My translation, from *TDn*, vol. 6: 232.
43. *Takamatsu Toyokichi den*: 82–93. For Takamatsu's strong connection with Shibusawa, see chapter 4, notes 52–55.
44. *Nakazawa Iwata kinenchō*: 72f. See also Kamatani 1989: 209.
45. Kamatani 1989: 286–297.
46. *Nakazawa Iwata kinenchō*: 67, 72 and 73–92.
47. Inoue 1932: 52–53 for the museum's perceived importance for the department.
48. Morris-Suzuki 1994: 82–83.
49. Morris-Suzuki 1994: 88–104.
50. On the separation of the Society of Chemical Industry of Japan from the Tokyo Chemical Society, see the analysis of Hirota Kōzō from his viewpoint of the conflict between applied-minded and pure-minded chemists within the society (Hirota 1988: 119–125). As I argue here, one should consider the transformation of the teaching of applied chemistry at Tokyo, overlooked by Hirota, and the increase of its nonchemical components as a major factor in this incident.
51. There has been no historical research on Kawakita, Tanaka, and twentieth-century developments of applied chemistry at Tokyo University. For obituary-like articles on them, see Tanaka 1961b and Nagai 1966.
52. *Kaki jisshū hōkoku*, *Kōjō keikaku*, and *Sotsugyō ronbun*. Tokyo Daigaku Kōgakubu Kagaku-Seimei-kei Toshoshitsu.
53. Furukawa 2010. Kita's wartime project of synthetic fuel production is discussed in Stranges 1993.

Epilogue

1 Gooday 2012 (as part of Bud 2012): 548–551.
2. *KDh, Sōsetsuhen*: 113. Publications about the history of Kyoto Imperial University and its chemistry-related departments include Kyoto Daigaku Rigakubu Kagaku Kyōshitsu sōritsu hyakushūnen kinen jigyōkai 1997, Kyoto Daigaku Kōgakubu Kagakukei hyakushunen kinen jigyō jikkō iinkai 1998, Furukawa 2010, and Furukawa 2012 in addition to *KDh*. Furukawa Yasu's entry of Sakurada Ichirō (Kita Gen-itsu's student) in *NDSB*, vol. 7: 330–335 is a useful English source on what he calls "the Kyoto school" of chemistry.
3. *KDh, Sōsetsuhen*: 119.
4. *KDh, Sōsetsuhen*: 126.
5. Amano 2009, vol. 2: 28.
6. *KDh, Bukyokushihen 1*: 478. The contemporary university calendar mentions two reasons: (1) to ensure interconnection between physics, pure chemistry, and mathematics; and (2) to enable some courses to take on more students. The latter implies low student enrollments. *Jpn. Calendar Kyoto Imp. Univ. 1904–1905*: 136–137.
7. *KDh, Sōsetsuhen*: 814–815.
8. *KDh, Bukyokushihen 1*: 812. See also *Jpn. Calendar Kyoto Imp. Univ. 1914–1915*: Appended campus map. Explanations of buildings are in the campus map appended to *Jpn. Calendar Kyoto Imp. Univ. 1916–1917.*
9. Fujii 1974; Gotō 1980b: 19–28.
10. Gotō 1980a: 21–22; Gotō 1980b: 14–16. See also Kamatani 1963: 18 and Kamatani 1994.
11. See the entry of Kainoshō written by myself in Miyachi et al. 2011–2013, vol. 1: 474. His collaboration with Kuhara on the Beckmann rearrangement started when he became a graduate student, and their first paper was published in 1907 as Kuhara and Kainoshō 1903–1908. *Jpn. Calendar Kyoto Imp. Univ. 1904–1905*: 31.
12. On Horiba, see Suito 1983. Kim 2005: 69–71 discussed the relationship between Horiba and his Korean student, Ree Taikyue.
13. Masao Horiba, transcript of interviews conducted by David C. Brock at HORIBA, Ltd. Kyoto, Japan, on November 19 and 20, 2004: 8–9. Oral History Collections, Othmar Library, Chemical Heritage Foundation.
14. Furukawa 2010: 6. See Kikuchi 2011: 239–240 for Sakurai's role in the establishment of the Riken.
15. On Fukui, see Bartholomew 1994 and James Bartholomew's entry of Fukui in *NDSB*, vol. 3: 85–89.
16. Furukawa 2010: 7.
17. Furukawa 2010: 3.

18. Furukawa 2010: 11 and 14 (n. 40). For Walker at MIT, see Servos 1980, Lécuyer 1998, and Lécuyer 1999.
19. Kikuchi 2011: 147 (n. 49).
20. Publications on the history of Tōhoku Imperial University and its Departments of Chemistry include Tōhoku Daigaku 1960, Tōhoku Daigaku hyakunenshi henshū iinkai 2003–2010, Tōhoku Kagaku Dōsōkai 1992, Kaji 2011a, and Kaji 2011b. Kaji Maranori's entry of Nozoe Tetsuo (Majima's student) in *NDSB*, vol. 5: 287–293 is a useful English source on Majima's research school.
21. Chapter 5, note 54.
22. Kaji 2011a: 173; Kaji 2011b: 189. Probably we would have to add Kuhara as well to the equation, but Kaji's assertion that the formation of a research school in organic chemistry did not originate at Tokyo Imperial University is correct.
23. See my entry of his student, Mizushima San-ichirō in *NDSB*, vol. 5, 167–171 as well as Tamamushi 1978, a recollection of another student of Katayama, Tamamushi Bun-ichi. Both of them were trained at Tokyo Imperial University after Katayama moved there in 1919 as a successor to Sakurai. Equally promising but tragic among Katayama's students was Yamada Nobuo, who graduated from Tōhoku Imperial University in 1919 (*Jpn. Calendar Tohoku Imp. Univ. 1919–1920*: 419). He published a paper on the electrochemistry of inner salt (3-methylsulfanilic acid) with Katayama as the lead author (Katayama and Yamada 1920), and another on the structural elucidation and chemical kinetics of diastase with an assistant professor at Tohoku, Yamazaki Eiichi, that same year (Yamazaki and Yamada 1920). Yamada later became a student of Marie Curie and worked with her daughter, Irène, at the Radium Institute in Paris. He died, most probably of radiation disease, at age 31. See Kawashima 2012.
24. Yoshihara 2003: 70.
25. Building plan to *Jpn. Calendar Tohoku Imp. Univ. Coll. Sci. 1913–1914*.
26. For the postwar orthodox understanding of the Japanese *kōza* system, see Cummings and Amano 1977. The term *kōza* remained in the Ministry of Education's University Establishment Standard (*daigaku setchi kijun*) until 2007.
27. Terasaki 2000b: 371–411, Amano 1977, and Amano 2009, vol. 1: 202–210.
28. Terasaki 2000b: 360–362 and Amano 2009: 205 and 209.
29. Amano 2009, vol. 1: 206.
30. Terasaki 2000b: 375 and 379 (n. 10).
31. Tokyo's Department of Physics came to occupy the old Main Science Building, originally designed as a "general chemical laboratory," when the Department of Chemistry was relocated to the new

Chemical Institute building in 1916. The former fell out of use by the Kanto Earthquake in 1923. See chapter 5, notes 63 and 65.

32. Terasaki 2000b: 371–379 gives a useful summary of such criticism up to 1973, when it was originally published right after the "age of activism." Comments from the US Scientific Advisory Group on the *kōza* system are in National Academy of Sciences Scientific Advisory Group 1947: 37. Critical comments by historians include Tsukahara 1978a: 20 and Dower 1993: 66.

33. See more recent and nuanced comments on the *kōza* system such as Uchida et al. 1990 and Araiso 1997.

Bibliography

Archival Sources

Bedford Museum, Bedford, Bedfordshire, UK.

Berlin-Brandenburgische Akademie der Wissenschaften, Berlin, Germany: Akademiearchiv, *Nachlass Ostwald.*

Chemical Heritage Foundation, Philadelphia, Pennsylvania, USA: Othmar Library Oral History Collections.

Columbia University Libraries, New York City, New York, USA: Rare Book and Manuscript Library.

Family Records Centre, London, UK.

Imperial College Archives, London, UK.

Ishikawa-ken Rekishi Hakubutsukan, Kanazawa, Japan: *Sakurai Jōji Kankei Shoshiryō.*

John Rylands University Library of Manchester, Manchester, UK: University of Manchester Archives.

Kokuritsu Kōbunshokan, Tokyo, Japan: *Kōbunroku; Kōbun fuzoku no zu.*

Kokuritsu Kokkai Toshokan, Tokyo, Japan: Kensei Shiryōshitsu, *Inoue Kaoru Monjo.*

Leeds University Library Special Collections, Leeds, UK.

Milton S. Eisenhower Library Special Collections, Baltimore, Maryland, USA: Johns Hopkins University Archives.

National Archives, Kew, Surrey, UK: Foreign Office Files.

Oberlin College Archives, Oberlin, Ohio, USA: Chemistry Department, College of Arts and Sciences (9/10), Series IV: Files related to select faculty.

Osaka Daigaku Fuzoku Toshokan, Toyonaka, Japan: *Majima Bunko.*

Royal Society, London, UK: Certificates of Election and Candidature.

Royal Society of Arts Archives, London, UK.

Royal Society of Chemistry, London, UK: Certificates of Candidates for Election, The Chemical Society

Rutgers University Library, New Brunswick, New Jersey, USA: William Elliot Griffis Collection; Rutgers University Archives.

Tokyo Daigaku Kōgakubu Kagaku- Seimei-kei Toshoshitsu, Tokyo, Japan.

Tokyo Daigaku Rigakubu Kagakuka, Tokyo, Japan.
Tokyo Daigaku Rigakubu Jimubu Shomu kakari, Tokyo, Japan.
Tokyo Daigaku Sōgō Toshokan, Tokyo, Japan.
Tokyo Daigakushi Shiryōshitsu, Tokyo, Japan: *Monbushō Ōfuku; Ryūgakusei Kankei Shorui; Watanabe Hiromoto Monjo.*
Tokyo Daigaku Sōgō Toshokan, Tokyo, Japan.
University College London Chemistry Department, London, UK.
University College London Library Services, Special Collections, London: Collection of notes and correspondence about A. W. Williamson (MS ADD 356); College Correspondence.
University College London Records Office, London, UK: University College London Council Minutes; University College London Senate Minutes; University College London, Register of Students.
Wellcome Library, Manuscripts and Archives Collection, London, UK: Lecture Notes Collection of William Dobinson Halliburton.

Other Primary and Secondary Sources

(A number in parentheses shows the year the item was originally published)

1875. *A Classified List of the English Books in the Tokio-Kaisei-Gakko.* Tokyo.
1881. *An Account of the New North Wing and Recent Additions to University College, London, opened February 16th, 1881, by the Rt. Hon. the Earl of Kimberley.* London: Dryden Press, J. Davy and Sons.
1884. *Second Report of the Royal Commissioners on Technical Instruction.* 5 vols. London: Printed by Eyre and Spottiswoode.
1888. Rigaku hakushi Matsui Naokichi kun shōden. In *Nihon hakushi zenden*, ed. Ogiwara Zentarō, pp. 104–106. Tokyo: Yoshioka shoseki.
1888. Rika Daigaku Kyōshitsu Rakusei su. In *Kanpō*, no. 1649 (December 26, 1888): 278.
1895. *Ranten seizō shiken hōkoku.* Tokyo: Nōshōmushō Shōkō-kyoku.
1928. *Nihon kagaku kaishi hōbun sōmokuji narabini sakuin.* Appended to *Nihon kagaku kaishi* 49.
1942. *Tokyo teikoku daigaku gakujutsu taikan: Rigakubu, Tokyo Tenmondai, Jishin kenkyūjo.* Tokyo: Tokyo Teikoku Daigaku.
1944. *Tokyo teikoku daigaku gakujutsu taikan: Kōgakubu, Kōkū Kenkyūjo.* Tokyo: Tokyo Teikoku Daigaku.
1956. *Ikeda Kikunae hakushi tsuiokuroku.* Tokyo: Ikeda Kikunae hakushi tsuiokukai.
1973. *Science Indicators 1972: Report of the National Science Board.* Washington, DC: National Science Board, National Science Foundation.
1989. *Shoku nihongi ichi. Shin nihon koten bungaku taikei 12.* Tokyo: Iwanami Shoten.
1990. *Shoku nihongi ni. Shin nihon koten bungaku taikei 13.* Tokyo: Iwanami Shoten.

2005. *Tsuyama han-i Kuhara-ke to kagakusha Kuhara Mitsuru: Tsuyama Yōgaku Shiryōkan heisei 17 nendo tokubetuten zuroku*. Tsuyama: Tsuyama Yōgaku Shiryōkan.

N. d. *Course of Laboratory Instruction in Chemistry. 3rd Year Class. 1st Term.* University of Tokio, Japan.

N. d. *Yamaguchi Hakushi Kenchiku Zushū.*

Allen, George C. 1962. *A Short Economic History of Modern Japan, 1867–1937: With a Supplementary Chapter on Economic Recovery and Expansion 1945–1960.* 4th edition. London: George Allen and Unwin.

Amano, Ikuo. 1977. Nihon no akademikku purofesshon: Teikoku daigaku ni okeru kyōju shūdan no keisei to kōza sei. *Daigaku kenkyū nōto* no. 30.

———. 2005 (1992). *Gakureki no shakaishi: Kyōiku to Nihon no kindai.* Tokyo: Heibonsha.

———. 2009. *Daigaku no Tanjō.* 2 vols. Tokyo: Chūō Kōronsha.

Angulo, A. J. 2012. The Polytechnic Comes to America: How French Approaches to Science Instruction Influenced Mid-nineteenth Century American Higher Education. *Hist. Sci.* 50: 315–338.

Araiso, Tsunehisa. 1997. Kyōin ninki sei to kyōiku kenkyū kasseika. *Kenkyū gijutsu keikaku gakkai nenji gakujutsu taikai kōen yōshishū* 12: 8–9.

Atkinson, Robert William. 1872. An Examination of the Recent Attack upon the Atomic Theory. *Philosophical Magazine*, 4th ser. 43: 428–433.

———. 1877. Tokyo fuka yōsui shiken-setsu. *Gs* 1, no. 5: 23–45.

———. 1878. The Water Supply of Tokio. *TASJ* 6: 87–98.

———. 1879a. Analysis of Surface Waters in Tokio. *TASJ* 7: 309.

———. 1879b. Kagaku rijitsu ai matsu no setsu. *Gs* 4: 131–172.

———. 1880. On Perthiocyanate of Silver. *JCST* 37: 226–232.

———. 1881a. The Chemistry of Sake-brewing. *Memoirs of the Science Department, Tokio Daigaku*. no. 6.

———. 1881b. Brewing in Japan. *CN* 44: 230–233.

———. 1881c. Nihon Jōshu hen. *Rika kaisui* 5.

Barker, George F. 1870. *A Textbook of Elementary Chemistry: Theoretical and Inorganic.* Louisville, K.Y.: John P. Morton.

Barff, Frederick Settle. 1878–1879. The Treatment of Iron to Prevent Corrosion. *JSA* 27: 390–397.

Bartholomew, James R. 1974. Japanese Culture and the Problem of Modern Science. In *Science and Values: Patterns of Tradition and Change*, eds. Arnold Thackray and Everett Mendelsohn, pp. 109–55. New York: Humanities Press.

———. 1989. *The Formation of Science in Japan: Building a Research Tradition.* New Haven, CT: Yale University Press.

———. 1994. Perspectives on Science and Technology in Japan: The Career of Fukui Ken'ichi. *Historia Scientiarum* 4: 47–54.

Beauchamp, Edward R. 1976. *An American Teacher in Early Meiji Japan.* Honolulu: University Press of Hawaii.

Beckmann, Ernst. 1889. Studien zur Praxis der Bestimmung des Molekulargewichts aus Dampfdruckerniedrigungen. *Zeitschrift für physikalische Chemie* 4: 532–552.

Bellot, H. Hale. 1929. *University College London 1826–1926*. London: University of London Press.

Berthelot, Marcellin. 1879. *Essai de mécanique fondée sur la thermochimie*. 2 vols. Paris: Dunod.

Berzelius, J. J. 1830. Ueber die Zusammensetzung der Weinsäure und Traubensäure (John's Säure aus den Vogesen), über das Atomengewicht des Bleioxyds, nebst allgemeinen Bemerkungen über solche Körper, die gleiche Zusammensetzung, aber ungleiche Eigenschaften besitzen. *Annalen der Physik und Chemie* 7: 305-335.

Beukers, H., L. Blussé, and R. Eggink, eds. 1987. *Leraar onder de Japanners: Brieven van dr. K.W. Gratama betreffende zijn verblijf in Japan, 1866–1871*. Amsterdam: De Bataafsche Leeuw.

Beukers, H., A. M. Luyendijk-Elshout, M. E. van Opstall, and F. Vos., eds. 1991. *Red-hair Medicine: Dutch-Japanese Medical Relations*. Amsterdam and Atlanta, Ga.: Ropogi.

Bowman, James E. (ed. Charles L. Broxam). 1866. *An Introduction to Practical Chemistry*. 5th edition. London: J. Churchill.

Bracken, Gregory. 2013. *The Shanghai Alleyway House: A Vanishing Urban Vernacular*. Abingdon, Oxon, and New York: Routledge.

Brinkley, F. 1901. *Japan: Its History Arts and Literature*, vol. 8: Keramic Art. Boston, MA, and Tokyo: J. B. Millet.

Brock, William H., ed. 1967. *The Atomic Debates*. Leicester: Leicester University Press.

———. 1981. The Japanese Connexion: Engineering in Tokyo, London, and Glasgow at the End of the 19th Century. *BJHS* 14: 227–243.

———. 1989. Building England's First Technical College: The Laboratories of Finsbury Technical College, 1878–1926. In *The Development of the Laboratory: Essays on the Place of Experiment in Industrial Civilization*, ed. Frank A. J. L. James, pp. 155–170. Basingstoke: Macmillan Press Scientific & Medical.

———. 1993. *The Norton History of Chemistry*. New York: Norton Press.

———. 1997. *Justus von Liebig: The Chemical Gatekeeper*. Cambridge: Cambridge University Press.

———. 1998. The Chemical Origins of Practical Physics. *Bulletin for the History of Chemistry* 21: 1–11.

———. 2011. *The Case of the Poisonous Socks: Tales from Chemistry*. Cambridge: Royal Society of Chemistry.

Brooke, John. 1991. *Science and Religion: Some Historical Perspectives*. Cambridge: Cambridge University Press.

Bud, Robert, ed. 2012. Focus: Applied Science. *Isis* 103 (3): 505–563.

Bud, Robert, and Gerrylynn K. Roberts. 1984. *Science versus Practice: Chemistry in Victorian Britain*. Manchester: Manchester University Press.

Campbell, W.A. 1998. Vapor Density, Boiling Point, and Freezing Point Apparatus. In *Instruments of Science: An Historical Encyclopedia*, eds. Deborah Jean Warner and Robert Bud, pp. 643–644. London and New York: The Science Museum, London and the National Museum of American History, Smithsonian Institution.

Chambers, David Wade, and Richard Gillespie. 2000. Locality in the History of Science: Colonial Science, Technoscience, and Indigenous Knowledge. *Osiris,* Second Series: 221–240.

Chapman, A. Chaston. 1910. Obituary Notices: Charles Graham. *Journal of the Chemical Society: Transactions* 97: 677–680.

Checkland, Olive. 1989. *Britain's Encounter with Meiji Japan, 1868–1912.* Basingstoke: Macmillan.

Chemical Society of Japan. 2003. *The Chemical Society of Japan: A 125-Year Quest for Excellence, 1878–2003.* Tokyo: The Chemical Society of Japan.

Chikashige, Masumi. 1895. Mercury Perchlorates. *JCST* 67: 1013–1017.

———. 1897. The Atomic Weight of Japanese Tellurium. *JCSIUTJ* 9: 123–128.

Cobbing, Andrew. 1998. *The Japanese Discovery of Victorian Britain: Early Travel Encounters in the Far West.* Richmond, Surrey: Japan Library.

———. 2000. *The Satsuma Students in Britain: Japan's Early Search for the 'Essence of the West.'* Richmond, Surrey: Japan Library.

Crookes, William, and Ernst Röhrig. 1869. *A Practical Treatise of Metallurgy, Adapted from the Last German Edition of Professor Kerl's Metallurgy.* 3 vols. London: Longman, Green.

Cummings, William K., and Ikuo Amano. 1977. The Changing Role of the Japanese Professor. *Higher Education* 6: 209–234.

Debus, Allen G. 1985. The Significance of Chemical History. *Ambix* 32: 1–14.

Demarest, William H. S. 1924. *A History of Rutgers College, 1766–1924.* New Brunswick, N.J.: Rutgers College.

Divers, Edward, and Seihachi Hada. 1899. Ethyl Ammonium Salenite and Non-existence of Amidoselenites. *JCST* 75: 537–541.

Divers, Edward, and Tamemasa Haga. 1884. On Hyponitrites. *JCST* 45: 78–87.

———. 1896. Amidosulphonic Acid. *JCST* 69: 1634–1653.

Divers, Edward, and Masataka Ogawa. 1899. Ethyl Ammoniumsulphite. *JCST* 75: 533–537.

———. 1900a. Ammonium Amidosulphite. *JCST* 77: 327–335.

———. 1900b. Products of Heating Ammonium Sulphites, Thiosulphate, and Trithionate. *JCST* 77: 335–340.

———. 1901. Ammonium and Other Imidosulphites. *JCST* 79: 1099–1103.

———. 1902. Preparation of Sulphamide from Ammonium Amido-sulphite. *JCST* 81: 504–507.

Divers, Edward, and M. Shimosé. 1883a. Lead-Chamber Deposit from the Brimstone Sulphuric Acid of Japan. *CN* 48: 283–284.

———. 1883b. On Tellurium Sulphoxide. *JCST* 43: 323–331.

Doak, Kevin M. 2007. *A History of Nationalism in Modern Japan: Placing the People*. Leiden: Brill.

Donnelly, James F. 1987. Chemical Education and the Chemical Industry in England from the Mid-Nineteenth to the Early Twentieth Century. Ph.D. dissertation, University of Leeds.

———. 1996. Defining the Industrial Chemist in the United Kingdom, 1850–1921. *Journal of Social History* 29: 779–796.

———. 1997. Getting Technical: The Vicissitudes of Academic Industrial Chemistry in Nineteenth-Century Britain. *History of Education* 26: 125–143.

Dore, Ronald. 1965. *Education in Tokugawa Japan*. London: Routledge & Kegan Paul.

Dower, John W. 1993. *Japan in War & Peace: Selected Essays*. New York: New Press.

Duke, Benjamin. 2009. *The History of Modern Japanese Education: Constructing the National School System, 1872–1890*. New Brunswick, NJ, and London: Rutgers University Press.

Eliot, Charles W., and Frank H. Storer. 1871. *A Manual of Inorganic Chemistry, Arranged to Facilitate the Experimental Demonstration of the Facts and Principles of the Science*. New York: Ivison, Blakeman, Taylor.

Elkana, Yehuda, Joshua Lederberg, Robert K. Merton, Arnold Thackray, and Harriet Zuckerman, eds. 1978. *Toward a Metric of Science: The Advent of Science Indicators*. New York, Chichester, Brisbane, and Toronto: John Wiley.

Elman, Benjamin. 2008. *A Cultural History of Modern Science in China*. Cambridge, MA. and London: Harvard University Press.

Fan, Fa-ti. 2004. *British Naturalists in Qing China: Science, Empire, and Cultural Encounter*. Cambridge, MA: Harvard University Press.

Foster, George Carey. 1905. Alexander William Williamson. *Journal of the Chemical Society: Transactions* 87: 605–618.

Foucault, Michel (trans. Alan Sheridan). 1979. *Discipline and Punish: The Birth of the Prison*. Harmondsworth: Penguin.

Fruton, Joseph S. 1985. *A Supplement to A Bio-Biography of the Biochemical Sciences Since 1800*. Philadelphia, PA: American Philosophical Society.

Fujii, Kiyohisa. 1974. Mitsuru KUHARA's studies on Beckmann rearrangement (in Japanese). *Kagakushi* no. 2: 11–15.

———. 1975. Atomism in Japan, 1868–1888. *JSHS* 14: 141–56.

Fujii, Yasushi. 1989. Yamao Yōzō to Yunibāshiti Karejji. *Eigakushi Kenkyū* 22: 77–89.

Fujimori, Terunobu. 2004 (1982). *Meiji no Tokyo keikaku*. Tokyo: Iwanami Shoten.

Furukawa, Yasu. 1996. From Chemistry to History: Historians of Chemistry and Their Community in Japan. *Historia Scientiarum* 6: 87–107.

———. 2010. Kita Gen-itsu and the Kyoto School's Formation (in Japanese). *Kagakushi* 37: 1–16.

———. 2012. From Fiber Chemistry to Polymer Chemistry: Ichiro Sakurada and the Development of the Kyoto School. *Kagakushi* 39: 1–40.

Galison, Peter. 1997. *Image and Logic: A Material Culture of Microphysics.* Chicago: University of Chicago Press.

Galison, Peter, and Emily Thompson, eds. 1999. *The Architecture of Science.* Cambridge, MA and London: MIT Press.

Gay, Hannah. 2000. 'Pillars of the College': Assistants at The Royal College of Chemistry, 1846–1871. *Ambix* 47: 135–169.

Geiger, Roger L. 1986. *To Advance Knowledge: The Growth of American Research Universities, 1900–1940.* New York and Oxford: Oxford University Press.

Gooday, Graeme J. N. 1989. Precision Measurement and the Genesis of Physics Teaching Laboratories in Victorian Britain. Ph.D. dissertation, University of Kent (Canterbury).

———. 1990. Precision Measurement and the Genesis of Physics Teaching Laboratories in Victorian Britain. *BJHS* 23: 25–51.

———. 1991. Teaching Telegraphy and Electrotechnics in the Physics Laboratory: William Ayrton and the Creation of an Academic Space for Electrical Engineering in Britain, 1873–1884. *History of Technology* 13: 73–111.

———. 2012. "Vague and Artificial": The Historically Elusive Distinction between Pure and Applied Science. *Isis* 103: 546–554.

Gooday, Graeme J. N., and Morris Low. 1998. Technology Transfer and Cultural Exchange: Western Scientists and Engineers Encounter Late Tokugawa and Meiji Japan. *Osiris*, Second Series 13: 99–128.

Gordon, Andrew. 2009 (2003). *A Modern History of Japan: From Tokugawa Times to the Present.* 2d edition. New York: Oxford University Press.

Gorman, Michael E. 2010. *Trading Zones and Interactional Expertise: Creating New Kinds of Collaboration.* Cambridge, MA and London: MIT Press.

Gotō, Ryōzō. 1980a. Kuhara Mitsuru no keireki to gyōseki (I). *Kagakushi* no. 12: 20–30.

———. 1980b. Kuhara Mitsuru no keireki to gyōseki (II). *Kagakushi* no. 13: 14–36.

Gotō, Sumio. 1985. Employment of David Murray as Superintendent of Education in Japan, I (in Japanese). *Kyōikugaku Zasshi: Nihon Daigaku Kyōiku Gakkai Kiyō*, no. 19: 18–30.

———. 1986. Employment of David Murray as Superintendent of Education in Japan, II (in Japanese). *Kyōikugaku Zasshi: Nihon Daigaku Kyōiku Gakkai Kiyō* no. 20: 1-15.

Gospel, Howard F., ed. 1991. *Industrial Training and Technological Innovation.* London and New York: Routledge.

Graham, Charles. 1873–1874. Cantor Lecture: On the Chemistry of Brewing. *JSA* 22: 188–194, 221–226, 245–252, 278–283, 297–302, 333–338, 366–373.

———. 1878–1879. Technical Education. *JSA* 27: 997–1000.

———. 1879–1880. Cantor Lectures: The chemistry of Bread-Making. *JSA* 28: 89–96, 97–103, 112–119, 123–129, 143–150.

Griffis, William Elliot. 1886. *The Rutgers Graduates in Japan: An Address delivered in Kirkpatrick Chapel, Rutgers College, June 16, 1885*. Albany, NY: Weed, Parsons.

———. 1900. *Verbeck of Japan: A Citizen of No Country*. New York: Fleming H. Revell.

———. 1924. Pioneering in Chemistry in Japan. *Industrial and Engineering Chemistry* 16: 1195–96.

Hada, Seihachi. 1895. Mercury and Bismuth Hypophosphites. *JCST* 67: 227.

———. 1896. How Mercurous and Mercuric Salts Change into Each Other. *JCST* 69: 1667–1673.

———. 1897. How Mercurous and Mercuric Salts Change into Each Other. *JCSIUTJ* 9: 161–194.

Haga, Tamemasa. 1912. Daibāsu sensei no den. *Tkk* 33: 1–26.

Haga, Tamemasa, and R. Majima. 1903. Über einige Anhydrobasen aus Diaminen der Fettreihe. *JCSIUTJ* 19: Article 7.

Haga, Tamemasa, and Yūkichi Ōsaka. 1895. The Acidimetry of Hydrogen Fluoride. *JCSIUTJ* 7: 255–267.

Hall, Ivan Parker. 1973. *Mori Arinori*. Cambridge, MA: Harvard University Press.

Hannaway, Owen. 1976. The German Model of Chemical Education in America: Ira Remsen at Johns Hopkins (1876–1913). *Ambix* 23: 145–164.

———. 1986. Laboratory Design and the Aim of Science: Andreas Libavius and Tycho Brahe. *Isis* 77: 584–610.

Hara, Heizō, 1992. *Bakumatsu yōgakushi no kenkyū*. Tokyo: Shin Jinbutsu Ōraisha.

Harris, J., and William H. Brock. 1974. From Giessen to Gower Street: Towards a Biography of Alexander William Williamson (1824–1904). *Ann. Sci.* 31: 95–130.

Harte, Negley, and John North. 1991. *The World of UCL, 1828–1990*. Revised edition. London: University College London.

Harwood, Jonathan. 2005. *Technology's Dilemma: Agricultural Colleges between Science and Practice in Germany, 1860–1934*. Bern: Peter Lang.

Hashimoto, Takehiko. 1993. New Historical Studies on Experiments and the Laboratory (in Japanese). *Kagakushi* 20: 107–121.

Hawkins, Hugh. 2002. *Pioneer: A History of the Johns Hopkins University, 1874–1889*. Ithaca, NY: Cornell University Press.

Heering, Peter, and Roland Wittje. 2011. *Learning by Doing: Experiments and Instruments in the History of Science Teaching*. Stutgart: Franz Steiner Verlag.

Helbig, Frances Yeomans. 1966. William Elliot Griffis: Entrepreneur of Ideas. M.A. thesis, University of Rochester, NY.

Henke, Christopher R., and Thomas F. Gieryn. 2008. Sites of Scientific Practice: The Enduring Importance of Place. In *The Handbook of Science and Technology Studies,* Third edition, eds. Edward J. Hackett, Olga Amsterdamska, Michael Lynch, and Judy Wajcman, pp. 353–376. Cambridge, MA, and London: MIT Press.

Hentschel, Klaus. 2002. *Mapping the Spectrum: Techniques of Visual Representation in Research and Teaching.* Oxford and New York: Oxford University Press.

Herzfeld, Michael. 2002. The Absence Presence: Discourses of Crypto-Colonialism. *The South Atlantic Quarterly* 101: 899–926.

Hirota, Kōzō. 1988. *Meiji no kagakusha: Sono kōsō to kujū.* Tokyo: Tokyo Kagaku Dōjin.

———. 1994. *Kagakusha Ikeda Kikunae: Sōseki, umami, doitsu.* Tokyo: Tokyo Kagaku Dōjin.

Hirschmeier, Johannes. 1964. *The Origins of Entrepreneurship in Meiji Japan.* Cambridge, MA: Harvard University Press.

Hörmann, Raphael, and Gesa Mackenthun, eds. 2010. *Human Bondage in the Cultural Contact Zone: Transdisciplinary Perspectives on Slavery and its Discourses.* Münster and New York: Waxmann.

Hofmann, A.W. 1866. *The Chemical Laboratories in Course of Erection in the Universities of Bonn and Berlin: Report Addressed to the Right Honourable the Lords of the Committee of Her Majesty's Most Honourable Privy Council of Education.* London: W. Clowes.

Holmes, Frederic L. 1989. The Complementarity of Teaching and Research in Liebig's Laboratory. *Osiris,* 2d series 5: 121–164.

Homburg, Ernst. 1999. The Rise of Analytical Chemistry and Its Consequences for the Development of the German Chemical Profession (1780–1860). *Ambix* 46: 1–32.

Horikoshi, Saburō. 1973 (1929). *Meiji shoki no Yōfū kenchiku.* Tokyo: Nanyōdō Shoten.

Hozumi, Shigeyuki. 1988. *Meiji ichi hōgakusha no shuppatsu: Hozumi Nobushige o megutte.* Tokyo: Iwanami Shoten.

Ikeda, Kikunae. 1890. Capillary Attraction in Relation to Chemical Composition, on the Basis of R. Schiff's Data. *JCSIUTJ* 3: 241–268.

———. 1898a. Yōbai no kagakuteki seishitsu to yōso yōeki no iro tono kankei ni tukite. *Tkk* 19: 120–142.

———. 1898b. Shintō atsuryoku no honshō ni tukite. *Tkk* 19: 1010–1030.

———. 1899. Bōchōkei o mochiite kasuibunkai no sokudo o sokutei suru hōhō. *Tkk* 20: 471–480.

———. 1902. Seihin kagaku to butsuri kagaku tono kankei ni tukite. *Tkk* 23: 1145–1154.

———. 1904. Saikin nijūgo nenkan honpō kagaku no hattatsu. *Tkk* 24: 470–487.

———. 1909. Shin chōmiryō ni tuite. *Tkk* 30: 820–836.

———. 1915. Haga Tamemasa shi shōden. *Tkk* 36: 11–25.

———. 1956 (1933). Ajinomoto hatsumei no dōki. In *Tsuiokuroku* (1956): 223–225.

——— (trans. Yoko Ogiwara and Yuzo Ninomiya). 2002. New Seasonings. *Chemical Senses* 27: 847–849.

Ikeda, Kikunae, and Kametaka Tokuhei. 1899. Dōzokutai no futten ni tukite. *Tkk* 20: 5–41.

Iinuma, Kazumasa, and Sugano Tomio. 2000. *Takamine Jōkichi no shōgai: Adorenarin hakken no shinjitsu*. Tokyo: Asahi Shinbunsha.

Imai, Ichiryō. 1994. *Ōzubon kikō: Samurai no musume to musubareta eijin ikka o otte*. Kanazawa: Hokkoku Shinbunsha.

Inada, Nobuji. 1984. Rosukō-shi kagaku. *Kk* 23: 129–39.

Inoue, Jinkichi. 1932. Takamatsu Hasushi to Kōka Daigaku Ōyō Kagaku Kyōshitsu. In *Takamatsu Toyokichi den* (1932): 49–55.

Inuzuka, Takaaki. 1974. *Satsuma-han Eikoku ryūgakusei*. Tokyo: Chūō Kōronsha.

———. 1986. *Mori Arinori*. Tokyo: Yoshikawa Kōbunkan.

———. 1987. *Meiji Ishin taigai kankei-shi kenkyū*. Tokyo: Yoshikawa Kōbunkan.

———. 2001. *Mikkō ryūgakusei tachi no Meiji Ishin: Inoue Kaoru to bakumatsu hanshi*. Tokyo: Nihon Hōsō Shuppan Kyōkai.

Inuzuka, Takaaki, and Ishiguro Keishō. 2006. *Meiji no wakaki gunzō: Mori Arinori kyōzō arubamu*. Tokyo: Heibonsha.

Ishizuki, Minoru. 1992 (1972). *Kindai Nihon no kaigai ryūgakushi*. Tokyo: Chūō Kōronsha.

Isono, Tokusaburō. 1979. *On Shoyu*. Tokyo: The Department of Science in Tokio Daigaku.

———. 1882. Shōyu o ronzu. *Tkk* 3: 1–11.

Jackson, Catherine M. 2006. Re-examining the Research School: August Wilhelm Hofmann and the Re-creation of a Liebigian Research School in London. *Hist. Sci.* 44: 281–319.

———. 2008a. Analysis and Synthesis in Nineteenth-Century Chemistry. Ph.D. dissertation, University of London (UCL).

———. 2008b. Visible Work: The Role of Students in the Creation of Liebig's Giessen Research School. *Notes & Records of the Royal Society* 62: 31–49.

Jansen, Marius B., ed. 1989. *The Cambridge History of Japan. Volume 5: The Nineteenth Century*. Cambridge: Cambridge University Press.

Johnson, Jeffrey A. 1989. Hierarchy and Creativity in Chemistry, 1871–1914. *Osiris*, 2d series. 5: 214–240.

———. 1992. Hofmann's Role in Reshaping the Academic-Industrial Alliance in German Chemistry. In *Die Allianz von Wissenschaft und Industrie: August Wilhelm Hofmann (1818–1892)*, eds. Christoph Meinel and Hartmut Scholz, pp. 167–182. Weinheim: VCH Verlagsgesellschaft.

Jones, Francis. 1872. *The Owens College Junior Course of Practical Chemistry.* London: Macmillan.

Jones, Hazel J. 1980. *Live Machines: Hired Foreigners and Meiji Japan.* Vancouver: University of British Columbia Press.

Kaiser, David. 2005a. *Drawing Theories Apart: The Dispersion of Feynman Diagrams in Postwar Physics.* Chicago: University of Chicago Press.

———, ed. 2005b. *Pedagogy and the Practice of Science: Historical and Contemporary Perspectives.* Cambridge, MA, and London: MIT Press.

Kaji, Masanori. 2011a. The Role of Riko Majima in the Formation of the Research Tradition of Organic Chemistry in Japan (in Japanese). *Kagakushi* 38: 173–185.

———. 2011b. Majima Toshiyuki to Nihon no yūki kagaku kenkyū dentō no keisei. In *Shōwa zenki no kagaku shisōshi*, ed. Kanamori Osamu, pp. 185–241. Tokyo: Keisō shobō.

Kakihara, Yasushi. 1996. Science versus Practice in Engineering Education of Modern Japan: Telegraphy at the Imperial College of Engineering, Tokyo (in Japanese). *Japan Journal for Science, Technology & Society* 5: 1–20.

———. 2002. Kōbushō no gijutsusha yōsei: Denshin no jirei o chūshin to shite. In *Kōbushō to sono Jidai*, ed. Suzuki Jun, pp. 57–82. Tokyo: Yamakawa Shuppansha.

Kamatani, Chikayoshi. 1963. The History of Research Organization in Japan. *JSHS* 2: 1-79.

———. 1988. *Gijutsu taikoku hyakunen no kei: Nihon no kindaika to kokuritsu kenkyū kikan.* Tokyo: Heibonsha.

———. 1989. *Nihon kindai kagaku kōgyō no seiritsu.* Tokyo: Asakura Shoten.

———. 1994. The Institute of Chemical Research [at Kyoto Imperial University] in its early years (in Japanese). *Kagakushi* 21: 1–37.

Kanao, Seizō. 1960. *Nagai Nagayoshi den.* Tokyo: Nihon Yakugakukai.

Karasawa, Tomitarō. 1990 (1974). *Kōshinsei: Bakumatsu Ishin-ki no eriito. Karasawa Tomitarō chosakushū 4.* Tokyo: Gyōsei, 1990.

Katayama, Masao. 1907. Über die Natur der Jodstärke. *Zeitschrift für anorganische Chemie* 56: 209–217.

———. 1950. Ichi ni no omoide to kansō. *Knr* 4: 63–67.

———. 1956. Ikeda Kikunae sensei no omoide. In *Tsuiokuroku* (1956): 37–44.

Katayama, Masao, and Yamada Nobuo. 1920. 3 mechiru surufaniru san to kyō denkaishitsu tono kankei ni tuite. *Tkk* 41: 193–224.

Kawashima, Keiko. 2012. Nobuo Yamada and Toshiko Yuasa: Two Japanese Scientists Who Became the Pupils of the Curies (in Japanese). *Kagakushi* 39: 179–190.

Key, T. Hewitt, and Alexander W. Williamson. 1858. *Invasion Invited by the Defenceless State of England.* London: Privately printed.

Kikuchi, Yoshiyuki. 1999. A List of the First Occupations of the Graduates of the Department of Chemistry, Tokyo (Imperial) University, 1877–1926 (in Japanese). *Kagakushi* 26: 102–112.

———. 2000. Redefining Academic Chemistry: Joji Sakurai and the Introduction of Physical Chemistry into Meiji Japan. *Historia Scientiarum* 9: 215–56.

———. 2004. Joji Sakurai and His Connections with British Chemists (in Japanese). *Kagakushi* 31: 239–267.

———. 2005. Wilhelm Ostwald and the Japanese Chemists. In *Wilhelm Ostwald at the Crossroads between Chemistry, Philosophy and Media Culture*, eds. Britta Görs, Nikos Psarros, and Paul Ziche, pp. 101–113. Leipzig: Leipziger Universitätsverlag.

———. 2009. Samurai Chemists, Charles Graham, and Alexander William Williamson at University College London, 1863–1872. *Ambix* 56: 115–137.

———. 2011. World War I, International Participation, and Reorganisation of the Japanese Chemical Community. *Ambix* 58: 136–149.

———. 2012. Cross-national Odyssey of a Chemist: Edward Divers at London, Galway, and Tokyo. *Hist. Sci.* 50: 289–314.

Kim, Dong-won. 2005. Two Chemists in Two Koreas. *Ambix* 52: 67–84.

Kinoshita, Naoyuki, Ōba Hideaki, and Kishida Shōgo. 2005. *Tokyo Daigaku Hongō kyanpasu annai*. Tokyo: Tokyo Daigaku Shuppankai.

Klein, Ursula. 2008. The Laboratory Challenge: Some Revisions of the Standard View of Early Modern Experimentation. *Isis* 99: 769–782.

Knight, David. 1992. *Ideas of Chemistry*. London: Athlone Press.

Knight, David, and Helge Kragh, eds. 1998. *The Making of a Chemist: The Social History of Chemistry in Europe, 1789–1914*. Cambridge and New York: Cambridge University Press.

Knight, Patrick. 2001. J & F Howard – Britannia Iron Works, Bedford. *Stationary Engine* no. 326: 24–26.

Koertge, Noretta, et al., eds. 2007. *New Dictionary of Scientific Biography*. 8 vols. Farmington Hills, Mich.: Scribner's/Thomson Gale.

Kōga, Yoshimasa. 1880. Kanda jōsui suishitsu bunseki-hyō fuki. *Tkk* 1: 166–168.

Kohler, Robert E., ed. 2008. Focus: Laboratory History. *Isis* 99 (4): 761–802.

Koizumi, Kenkichiro. 1973. The Development of Physics in Meiji Japan: 1868–1912. Ph.D. dissertation, University of Pennsylvania.

———. 1975. The Emergence of Japan's First Physicists: 1868–1900. *HSPS* 6: 3–108.

Kondō, Yutaka. 1999. *Meiji Shoki no Gi-Yōfū Kenchiku no Kenkyū*. Tokyo: Rikō Gakusha.

Kozai, Yoshinao. 1911. Rigaku hakushi Matsui Naokichi kun ryakuden. *Tkk* 32: 1–6.

Kuhara, Mitsuru. 1877. Ahen no gai. In *Kaisei Gakkō kōgishitsu hakkai enzetsu*, pp. 29-36. Tokyo: Tokyo Kaisei Gakkō.

———. 1879a. Tokyo fuka seisui bunseki setsu. *Gs* 4, no. 20: 196–208.

———. 1879b. Yumoto kōsen bunseki-hyō narabini iji kōyō. *Gs* 5, no. 24: 70–73.

———. 1879c. On the Red Colouring Matter of the Lithospermum Erythrorhizon. *JCST* 35: 22–25.

———. 1879–1880. A Method for Estimating Bismuth Volumetrically. *ACJ* 1: 326–329.

———. 1881–1882. Concerning Phtalimide. *ACJ* 3: 26–30.

———. 1882. Yūki kagaku no kōkyū. *Tkk* 3: 49–63.

Kuhara, Mitsuru, and Tadaka Kainoshō. On Beckmann's Rearrangement. *Memoirs of the College of Science and Engineering, Kyōto Imperial University* 1: 254-264.

Kumazawa, Eriko. 2004. *'Gakusei' izen ni okeru Kokumin Kaihei Shisō to Puroisen* Gakkō Kyōiku Seido eno Keitō. Heisei 14–15 nendo Kagaku Kenkyūhi Hojokin [Kiban Kenkyū (C) (2)] Kenkyū Seika *Hōkokusho (Kadai Bangō 14510309).*

Kurahara, Miyuki. 1987a. Oyatoi gaikokujin kyōshi W. E. Gurifisu no kyōiku katsudō: Kaisei Gakkō ni okeru jugyō o chūshin ni. *Kantō kyōiku gakkai kiyō* no. 14: 21–34.

———. 1987b. Oyatoi gaikokujin kyōshi no keiyaku to shogū: W. E. Gurifisu no baai. *Nihon kyōikushi kenkyū* no. 6: 1–22.

———. 1997. E. W. Clark's Chemistry and Physics Classes at Denshūjo-school of Shizuoka-han: From His Letters To William Elliot Griffis (in Japanese). *Musashigaoka Tanki Daigaku Kiyō* 5: 1–19.

———. 2000. W. E. Griffis and New Scientific Curriculum in Rutgers College (in Japanese). *Kk* 39: 144–152.

———. 2005. The Griffis' Journal of Tokyo Years (1873/4/1–1874/9/17) (No. 2). *Musashigaoka Tanki Daigaku Kiyō* 13: 23–57.

Kurasawa, Takashi. 1973. *Gakusei no kenkyū*. Tokyo: Kōdansha.

———. 1983. *Bakumatsu kyōikushi no kenkyū 1: Chokkatsu gakkō seisaku.* Tokyo: Kōdansha.

Kyoto Daigaku hyakunenshi henshū iinkai, ed. 1997–2001. *Kyoto Daigaku hyakunenshi*. 8 vols. Kyoto: Kyoto Daigaku kōenkai.

Kyoto Daigaku Rigakubu Kagaku Kyōshitsu sōritsu hyakushūnen kinen jigyōkai, ed. 1997. *Kyoto Daigaku Rigakubu Kagaku Kyōshitsu hyakunen no ayumi*. Kyoto Daigaku Rigakubu Kagaku Kyōshitsu sōritsu hyakushūnen kinen jigyōkai.

Kyoto Daigaku Kōgakubu Kagakukei hyakushunen kinen jigyō jikkō iinkai, 1998. *Dentō no keisei to keishō: Kyoto Daigaku Kōgakubu Kagakukei hyakunenshi*. Kyoto: Kyoto Daigaku Kōgakubu Kagakukei hyakushunen kinen jigyō jikkō iinkai.

Kyū Kōbu Daigakkō Shiryō Hensankai. 1931. *Kyū Kōbu Daigakkō Shiryō*. Tokyo: Toranomonkai.

Lahiri, Shompa. 1995. Metropolitan Encounters: A Study of Indian Students in Britain, 1880–1930. Ph.D. dissertation, University of London (SOAS).

Larson, Robert. 1950. Charles Frederick Chandler: His Life and Work. Ph.D. dissertation, Columbia University.

Lécuyer Christophe. 1998. Academic Science and Technology in the Service of Industry: MIT Creates a "Permeable" Engineering School. *American Economic Review* 88: 23–33.

———. 1999. MIT, Progressive Reform, and "Industrial Service," 1890–1920. *HSPS* 26: 1–54.

———. 2010. Patrons and A Plan. In *Becoming MIT: Moments of Decision*, ed. David Kaiser, pp. 59–80. Cambridge, MA: MIT Press.

Lightman, Bernard V. ed. 2004. *The Dictionary of Nineteenth-Century British Scientists.* 4 vols. Chicago: University of Chicago Press.

Lintsen, Harry, and Martijn Bakker. 1994. *Geschiedenis van de techniek in Nederland: De wording van een moderne samenleving 1880–1890. Deel V: Techniek, beroup en praktijk.* 's Gravenhage and Zutphen: Stichting Historie der Techniek and Walburg Pers.

Loew, Oscar. 1896. Physiological Action of Amidosulphonic Acid. *JCST* 69: 1662–1664.

Low, Morris. 2005. *Science and the Building of a New Japan.* New York and Basingstoke: Palgrave Macmillan.

Lundgreen, Peter. 1987. Fachschulen. In *Handbuch der deutschen Bildungsgeschichte, Band III 1800–1870: Von der Neuordnung Deutschlands bis zur Gründung des Deutschen Reiches*, eds. K.-E. Jeismann and P. Lundgreen, pp. 293–305. München: Verlag C. H. Beck.

———. 1990. Engineering Education in Europe and the U.S.A., 1750–1930: The Rise to Dominance of School Culture and the Engineering Profession. *Ann. Sci.* 47: 33–75.

Lundgren, Anders, and Bernadette Bensaude-Vincent, eds. 2000. *Communicating Chemistry: Textbooks and Their Audiences, 1789–1939.* Canton, MA: Science History Publications.

MacLeod, Roy. 2000. *Nature and Empire: Science and Colonial Enterprise. Osiris*, 2d series, vol. 15.

Majima, Toshiyuki. 1954. Waga shōgai no kaiko. *Knr* 8: 1–11 and 137–146.

Markus, Thomas A. 1993. *Buildings & Power: Freedom and Control in the Origin of Modern Building Types.* London and New York: Routledge.

Matheson, Hugh M. 1899. *Memorials of Hugh M. Matheson. Edited by His Wife. With a Prefatory Note by the Rev. J. Oswald Dykes.* London: Hodder & Stoughton.

Matsubara, Shunzō. 1899. Futten no jōshō ni yoreru bunshiryō no sokutei. *Tkk* 20: 591-613.

Matsui, Naokichi. 1880. Examination of the Raw Materials Used for the Arita Porcelain. *Journal of the American Chemical Society* 2: 315–332.

———. 1882. Genshiron enkaku no gairyaku. *Tkk* 3: 35–48.

McCaughey, Robert A. 2004. *Stand, Columbia: A History of Columbia University in the City of New York, 1754–2004.* New York: Columbia University Press.

Miller, Richard E. 1994. Fault Lines in the Contact Zone. *College English* 56: 389–408.

Miller, William Allen. 1867. *Elements of Chemistry: Theoretical and Practical*. Part I: Chemical Physics. 4th edition, with additions. London: Longmans, Green, Reader, and Dyer.

Mills, Edmund J., and Jokichi Takamine. 1883. On the Absorption of Weak Reagents by Cotton, Silk, and Wool. *JCST* 43: 142–153.

Miyachi, Masato. 2001. Gijutsu Kyōikushi. In *Sangyō gijutsushi*, eds. Nakaoka Tetsurō, Suzuki Jun, Tsutsumi Ichirō, and Miyachi Masato, pp. 433–448. Tokyo: Yamakawa Shuppansha.

Miyachi, Masato, Satō Yoshimaru, and Sakurai Ryōju, eds. 2011–2013. *Meiji jidaishi daijiten*. 4 vols. Tokyo: Yoshikawa Kōbunkan.

Miyamoto, Matarō. 1999. *Kigyōka tachi no chōsen*. Tokyo: Chūō Kōron Shinsha.

Miyoshi, Nobuhiro. 1979. *Nihon kōgyō kyōiku seiritsu shi no kenkyū: Kindai nihon no kōgyōka to kyōiku*. Tokyo: Kazama Shobō.

Mizuta, Masakichi. 1932. Takamatsu sensei to watashi. In *Takamatsu Toyokichi den* (1932): 425–438.

Mody, Cyrus C. M., and David Kaiser. 2008. Scientific Training and the Creation of Scientific Knowledge. In *The Handbook of Science and Technology Studies*, 3d edition, eds. Edward J. Hackett, Olga Amsterdamska, Michael Lynch, and Judy Wajcman, pp. 378–402. Cambridge, MA, and London: MIT Press.

Molhuysen, P.C., ed. 1911–1937. *Nieuw Nederlandsch Biografisch Woordenboek*. 10 vols. Leiden: A. W. Sijthoff.

Morishita, Masaaki. 2010. *The Empty Museum: Western Cultures and the Artistic field in Modern Japan*. Aldershot: Ashgage.

Moriya, Monoshirō. 1881. On Menthol or Peppermint Camphor. *JCST* 39: 77–83.

Morrell, Jack B. 1972. The Chemist Breeders: The Research Schools of Liebig and Thomas Thomson. *Ambix* 19: 1–46.

Morris-Suzuki, Tessa. 1994. *The Technological Transformation of Japan: From the Seventeenth to the Twenty-First Century*. Cambridge: Cambridge University Press.

———. 1995. The Great Translation: Traditional and Modern Science in Japan's Industrialisation. *Historia Scientiarum* 5: 103–16.

Murai, Masatoshi. 1915. *Shishaku Inoue Masaru kun shōden*. Tokyo: Inoue Shishaku Dōzō Kensetsu Dōshikai.

Muramatsu, Teijiro. 1954. Nihon no kōgaku sōseiki no jakkan no mondaiten: toku ni kōgakusha nit suite. *Kk* 32: 8–14.

Murase, Hisayo. 2003. Furubekki no haikei: Oranda, America no chōsa o chūshin ni. *Momoyama gakuin daigaku kirisutokyō ronshū* no. 39: 55–78.

Murray, David. 1917. Hatakeyama Yoshinari of Japan. *The Japanese Student* 1, no. 3: 108–113.

Nagai, Shōichirō. 1966. Ko Tanaka Yoshio sensei o shinobu. *Kagaku kyōiku* 14 (1966): 256–257.

Nakano, Minoru. 1995. Kōbu Daigakkō no ikan to tōgō. *Kanagawa Daigaku Hyōron* no. 22: 192–193.

———. 1999. *Tokyo Daigaku monogatari: Mada kimi ga wakakatta koro.* Tokyo: Yoshikawa Kōbunkan.

———. 2003. *Kindai Nihon Daigaku Seido no Seiritsu.* Tokyo: Yoshikawa Kōbunkan.

Nakaoka, Tetsurō. 2001. Sangyō Gijutsu to sono Rekichi. In *Sangyō Gijutsushi*, eds. Nakaoka Tetsurō, Suzuki Jun, Tsutsumi Ichirō, and Miyachi Masato, pp. 7-33. Tokyo: Yamakawa Shuppansha.

———. 2006. *Nihon kindai gijutsu no keisei: "Dentō" to "kindai" no daimamikkusu.* Tokyo: Asahi Shimbunsha.

Nakayama, Shigeru. 1978. *Teikoku daigaku no tanjō: Kokusai hikaku no naka deno Tōdai.* Tokyo: Chūō Kōronsha.

Nakazawa, Iwata (Nakasawa, I.). 1879. *The Chemistry of Copper Smelting.* Tokyo: The Department of Science, Tokio Daigaku.

———. 1882. Tōki seizō no ryakureki. *Tkk* 3: 90–94.

———. 1883. Hari yōsei no setsu. *Tkk* 4: 25–40.

———. 1888. Doitsu-koku jiki seizō no gairyaku: Fu honpō jiki seizō no kairyō kō. *Tkk* 9: 55–97.

———. 1889. Hari bin seizō no gairyaku. *Tkk* 10: 105–112.

———. 1892. Wagakuni ni okeru ryūsan oyobi sōda seizō ni tuite. *Tkk* 13: 21–27.

———. 1932. Ōyō kagakuka no sōsetsu. In *Takamatsu Toyokichi den* (1932): 35–39.

———. 1938. Waguneru sensei raireki dai ichi dan. In *Waguneru sensei tuikaishū*, ed. Otogorō Umeda, pp. 72–90. Tokyo: Ko Waguneru Hakushi Kinen Jigyōkai.

Nakazawa Iwata hakushi kiju shukuga kinenkai, ed. 1935. *Nakazawa Iwata hakushi kiju shukuga kinenchō.* Tokyo: Nakazawa Iwata Hakushi kiju shukuga kinenkai.

National Academy of Sciences Scientific Advisory Group. 1947. *Reorganization of Science and Technology in Japan: Report of the Scientific Advisory Group of the National Academy of Sciences United States of America.* Tokyo.

Nihon Bunseki Kagakukai, ed. 1981. *Nihon bunseki kagakushi.* Tokyo: Tokyo Kagaku Dōjin.

Nihon Bunseki Kagakukai sōritsu jusshūnen kinenshi henshū iinkai, ed. 1963. *Sōritsu jusshūnen kinenshi.* Tokyo: Nihon Bunseki Kagakukai.

Nihon Kagakukai, ed. 1978. *Nihon no kagaku hyakunenshi: Kagaku to kagaku kōgyō no ayumi.* Tokyo: Tokyo Kagaku Dōjin.

Nish, Ian, ed. 1998. *The Iwakura Mission in America and Europe: A New Assessment.* Richmond, Surrey: Japan Library.

Nishimura, Masamori. 1977. Hatakeyama Yoshinari yōkō nikki (Sugiura Kōzō seiyō yūgaku nisshi). *Sankō shoshi kenkyū* no. 15: 1–16.

Nishizawa, Yūshichi, 1923. Tōkyō Teidai konjaku monogatari. *Kagaku chishiki* 3: 106-108, 255–256, 410–411, 497–499, and 579–581.

Noyes, William A. 1927. Ira Remsen. *Science* 66: 243–246.

Numakura, Mitsuho. 1988. Masaki Taizō ryakuden. *Yamaguchi-ken chihōshi kenkyū* no. 59: 35–47.

Numata, Jirō. 1989. *Yōgaku. Nihon rekishi sōsho 40.* Tokyo: Yoshikawa Kōbunkan.

Ogawa, Ayako. 1998. *Bakumatsu Chōshū han yōgakushi no kenkyū.* Kyoto: Shibunkaku Shuppan.

Ogawa, Kazumasa, ed. 1900. *Imperial University of Tōkyō/Tokyo Teikoku Daigaku.* Tokyo: Ogawa shashin seihanjo.

Ōhashi, Akio, and Hirano Hideo. 1988. *Meiji ishin to aru oyatoi gaikokujin: Furubekki no shogai.* Tokyo: Shin Jinbutsu Ōraisha.

Ōhashi, Ryōsuke, ed. 1993. *Bunka no hon-yaku kanōsei: Kokusai Kōtō Kenkyūjo shinpojiumu.* Kyoto: Jinbun Shoin.

Oki, Hisaya. 2013. W. E. Griffis's Lecture Note, "Kagaku Hikki" (in Japanese). *Kagakushi* 39: 191–198.

Ōkubo, Toshiaki. 1944. *Mori Arinori.* Tokyo: Bunkyō Shoin.

———, ed. 1972. *Mori Arinori zenshū.* 3 vols. Tokyo: Senbundō Shoten.

———. 1976. *Meirokusha kō.* Tokyo: Rittaisha.

———. 1988. Bunmei kaika. In Ōkubo Toshiaki, *Meiji no shisō to bunka. Ōkubo Toshiaki rekishi chosakushū* 6, pp 107–146. Tokyo: Yoshikawa Kōbunkan.

Okuno, Hisateru. 1980. *Edo no kagaku.* Tokyo: Tamagawa Daigaku Shuppanbu.

Ōmachi, Keigetsu, and Ikari Shizan. 1986 (1924). *Sugiura Shigetake sensei.* Ōtsu: Sugiura Shigetake Sensei Kenshōkai.

Ono, Heihachirō, and Fujise Shin-ichirō. 1961. Ogawa Matasaka sensei. *Kagaku* (Chemistry) 16: 220–225.

Ortiz, Fernando. 1995 (1940). *Cuban Counterpoint: Tobacco and Sugar.* Durham, NC, and London: Duke University Press.

Ōsaka, Yūkichi. 1949. Kagakusha to shite no yo no omoide. *Knr* 3: 487–495 and 549–555.

Ostwald, Wilhelm. 1897. *Die wissenschaftlichen Grundlage der analytischen Chemie.* Zweite Auflage. Leipzig: W. Engelmann.

———. 1898. *Das physikalisch-chemische Institut der Universität Leipzig und die Feier seiner Eröffnung am 3. Januar 1898.* Leipzig: Wilhelm Engelmann.

——— (trans. Kametaka Tokuhei). 1899. *Osutowarudo-shi Bunseki Kagaku Genri.* Tokyo: Toyamabō.

Ōya, Shin-ichi, and Hara Masatoshi, eds. 1965. *Nkgt* 9: *Kyōiku* 2. Tokyo: Dai-ichi Hōki.

Ozawa, Takeshi, 2011. A German Science Teacher in Meiji Japan: G. Wagener's Upbringing and Education (in Japanese). *Kagakushi* 38: 28–36.

Picketts, Palmer C. 1930. *Rensselaer Polytechnic Institute: A Short History. Rensselaer Polytechnic Institute Engineering and Science Series No. 29.* Troy, NY.

Pickstone, John V. 2001. *Ways of Knowing: A New History of Science, Technology, and Medicine.* Chicago: University of Chicago Press.

Pratt, Mary Louise. 1992. *Imperial Eyes: Travel Writing and Transculturation.* London and New York: Routledge.

Principe, Lawrence. 2013. *The Secrets of Alchemy.* Chicago and London: University of Chicago Press.

Prown, Jules David. 1993. The Truth of Material Culture: History or Fiction? In *History from Things: Essays on Material Culture*, eds. Steven Lubar and W. David Kingery, pp. 1–19. Washington, DC, and London: Smithsonian Institution Press.

Ramberg, Peter J. 2003. *Chemical Structure, Spatial Arrangement: The Early History of Stereochemistry, 1874–1914.* Aldershot: Ashgate.

Reardon-Anderson, James. 1991. *The Study of Change: Chemistry in China, 1840–1949.* Cambridge: Cambridge University Press.

Remsen, Ira. 1923. Biographical Memoir Harmon Northrop Morse. *National Academy of Sciences Biographical Memoirs* 21, no. 11.

Remsen, Ira, and W. Burney. 1880–1881. On the Oxidation of Substitution Product of Aromatic Hydrocarbons. VII. – Sulphoterephthalic Acid. *ACJ* 2: 405–413.

Remsen, Ira, and C. Fahlberg. 1879–1880. On the Oxidation of Substitution Products of Aromatic Hydrocarbons. IV. – On the Oxidation of Orthololuenesulphamide. *ACJ* 1: 426–438.

Remsen, Ira, and M. Kuhara. 1880–1881. On the Oxidation of Substitution Product of Aromatic Hydrocarbons. VIII. – Sulphoterephthalic acid from paraxylene-sulphonic acid. *ACJ* 2: 413–416.

———. 1881–1882. On the Oxidation of Substitution Products of Aromatic Hydrocarbons. XL. – On the Conduct of Nitro-Meta-Xylene towards Oxidizing Agents. *ACJ* 3: 424–433.

Roberts, Gerrylynn K. 1976. The Establishment of the Royal College of Chemistry: An Investigation of the Social Context of Early Victorian Chemistry. *HSPS* 7: 437–485.

———. 1998. 'A Plea for Pure Science': The Ascendancy of Academia in the Making of the English Chemist, 1841–1914. In *The Making of the Chemist: The Social History of Chemistry in Europe, 1789–1914*, eds. David Knight and Helge Kragh, pp. 107–119. Cambridge: Cambridge University Press.

Robins, Edward Cookworthy. 1887. Technical School and College Building: Being a Treatise on the Design and Construction of Applied Science and Art Buildings, and their Suitable Fittings and Sanitation, with a Chapter on Technical Education. London: Whittaker.

Rocke, Alan J. 1984. *Chemical Atomism in the Nineteenth Century: From Dalton to Cannizzaro.* Columbus, Ohio: Ohio State University Press.

———. 2003. Origin and Spread of the 'Giessen Model' in University Science. *Ambix* 50: 90–115.

———. 2004. Williamson, Alexander William (1824–1904). In *Oxford Dictionary of National Biography*, vol. 59, pp. 339–341. Oxford: Oxford University Press.

———. 2010. *Image and Reality: Kekulé, Kopp, and the Scientific Imagination*. Chicago: University of Chicago Press.

Roden, Donald. 1980. *Schooldays in Imperial Japan: A Study in the Culture of a Student Elite*. Berkeley, and London: University of California Press.

Ronalds, Edmund, Thomas Richardson, and Henry Watts, eds. 1855–1867. *Chemical Technology; or Chemistry in its Applications to Arts and Manufactures*, vol. 1, parts 1–5. London: Baillière.

Roscoe, Henry E. 1868. *Lessons in Elementary Chemistry: Inorganic and Organic*. New York: W. M. Wood.

———. 1878. *Description of the Chemical Laboratories at the Owens College, Manchester*. [Manchester]: J. E. Cornish.

———. 1881. Presidential Address. In Society of Chemical Industry, *Proceedings of First General Meeting, held at the Institution of Civil Engineers, Great George Street, Westminster, June 28 and 29th, 1881*, pp. 3–7. Manchester: Printed by William T. Emmott.

———. 1887. *Record of Work done in the Chemical Department of the Owens College, 1857–1887*. London and New York: Macmillan.

Roscoe, Henry E., and Carl Schorlemmer. 1881–1892. *A Treatise on Chemistry, vol. III: The Chemistry of the Hydrocarbons and their Derivatives, or Organic Chemistry*. 6 Parts. London: Macmillan.

Rossiter, Margaret W. 1975. *The Emergence of Agricultural Science: Justus Liebig and the Americans, 1840–1880*. New Haven and London: Yale University Press.

———. 1977. The Charles F. Chandler Collection. *Technology and Culture* 18: 222–230.

Royal Society of London. 1867–1925. *Catalogue of Scientific Papers*. 19 vols. London: For her Majesty's Stationery Office.

Russell, Colin A. 1996. *Edward Frankland: Chemistry, Controversy, and Conspiracy in Victorian England*. Cambridge: Cambridge University Press.

Russell, Colin A., Noel G. Coley, and Gerrylynn K. Roberts. 1977. *Chemists by Profession: The Origins and Rise of the Royal Institute of Chemistry*. Milton Keynes: Open University Press.

Sakanoue, Masanobu. 1979. Sakurai Jōji hakushi to sono kankei sho shiryō. *Kagakushi* no. 11: 3–13.

———. 1997. Seiyō kindai kagaku no ishoku- ikusei-sha Sakurai Jōji. *Kagakushi* 24: 157–68.

Sakurai, Jōji. 1880a. On Metallic Compounds Containing Bivalent Hydrocarbon Radicals. Part I. *BAAS Report* 50: 504–505.

———. 1880b. On Metallic Compounds Containing Bivalent Hydrocarbon Radicals. Part I. *JCST* 37: 658–661.

———. 1882. Kagaku to butsurigaku tono kankei o ronzu. *Tgz* no. 11: 1–3.

———. 1883. Netsu kagaku ron. *Tkk* 4: 55–92; *Tgz* no. 23: 44–49, 77–83, and 101–105.

———. 1884. Kagaku jugyō hō. *Tgz* no. 31: 1–7.

———. 1885. Kagaku bankin no shinpo. *Tgz* no. 47: 202–205.

———. 1888. Rigakusha no kairaku. *Tgz* no. 84: 437-442.

———. 1892a. Determination of the Temperature of Steam Arising from Boiling Salt Solutions. *JCST* 61: 495–508.

———. 1892b. Modification of Beckmann's Boiling Point Method of Determining the Molecular Weights of Substances in Solution. *JCST* 61: 989–1002.

———. 1894. Constitution of Glycocine and its Derivatives. *Proceedings of the Chemical Society* 10: 90–94.

———. 1895a. Constitution of Glycocine. *Proceedings of the Chemical Society* 12: 38–39.

———. 1895b. Constitution of Glycocol and its Derivates. (Appendix: General Theory and Nomenclature of Amido-acids.) *JCSIUT* 7: 87–107.

———. 1896. Molecular Conductivity of Amidosulphonic Acid. *JCST* 69: 1654–1662.

———. 1899. Kokka to rigaku. *Tokyo Gakushi Kaiin zasshi* 21: 1–20.

———. 1932. Rikagaku Kenkyūjo no Setsuritsu. In *Takamatsu Toyokichi den* (1932): 301–309.

———. 1940. *Omoide no kazukazu: Danshaku Sakurai Jōji ikō*. Tokyo: Kyūwakai.

Samejima, Jitsusaburō. 1942. Kagakuka. In *Tokyo Teikoku Daigaku gakujutsu taikan: Rigakubu, Tokyo Tenmondai, Jishin Kenkyūjo*, pp. 122–136. Tokyo: Tokyo Teikoku Daigaku.

———. 1956. Tsuiokuki. In *Tsuiokuroku* (1956): 106–115.

Sand, Jordan. 2005. A Short History of MSG: Good Science, Bad Science, and Taste Cultures. *Gastronomica: The Journal of Food and Culture* 5: 38–49.

Sasaki, Yukiyoshi, and Tachibana Tarō. 1991. A List of the Thesis Defenses at the Department of Chemistry, Tokyo Imperial University, 1916–1953 (in Japanese). *Kagakushi* 18: 23–32 and 85–91.

Satō, Shōsuke. 1980. *Yōgakushi no kenkyū*. Tokyo: Chūō Kōronsha.

Schaffer, Simon, Lissa Roberts, Kapil Raj, and James Delbourgo, eds. 2009. *The Brokered World: Go-Between and Global Intelligence, 1770–1820*. Sagamore Beach, MA: Science History Publications.

Scholz, Hartmut. 1992. August Wilhelm Hofmann und die Reform der Chemikerausbildung an deutschen Hochschulen. In *Die Allianz von Wissenschaft und Industrie: August Wilhelm Hofmann (1818–1892)*, eds. Christoph Meinel and Hartmut Scholz, pp. 221–233. Weinheim: VCH Verlagsgesellschaft.

Schorlemmer, Carl. 1874. *A Manual of the Chemistry of the Carbon Compounds; or, Organic Chemistry*. London: Macmillan.

Schwantes, Robert S. 1955. *Japanese and Americans: A Century of Cultural Relations*. New York: Harper.

Secord, James A. 2004. Knowledge in Transit. *Isis* 95: 654–672.

Seeley, Bruce. 1993. Research, Engineering, and Science in American Engineering Colleges, 1900–1960. *Technology and Culture* 34: 344–386.

Sekiguchi, Naosuke. 2009. Meiji shonen no mombu gyōsei to Tsuji Shinji. *Shagakuken ronshū: The Waseda Journal of Social Sciences* 14: 153–163.

Servos, John W. 1980. The Industrial Relations of Science: Chemistry at MIT 1900–1939. *Isis* 71: 531–549.

———. 1990. *Physical Chemistry from Ostwald to Pauling: The Making of a Science in America*. Princeton, NJ Princeton University Press.

Seth, Suman. 2010. *Crafting the Quantum: Arnold Sommerfeld and the Practice of Theory, 1890–1926*. Cambridge, MA, and London: MIT Press.

Shiba, Tetsuo. 1993. Orandajin no mita bakumatsu Meiji no Nihon: Kagakusha Haratama shokanshū. Tokyo: Saikon Shuppan.

———. 1999. Gottofuriito Waguneru (1831–1892). *Wakō Jun-yaku jihō* 67, no. 3: 2–4.

———. 2000. Dutch Chemist Gratama and Chemistry in Japan. *Historia scientiarum* 9: 181–190.

———. 2004. Takamatsu Toyokichi (1852–1937). *Wakō Jun-yaku jihō* 72, no. 3: 26–27.

———. 2006. *Nihon no kagaku no kaitakusha tachi*. Tokyo: Shōkabō.

Shibata, Yūji. 1939. Sakurai Sensei no tsuioku. *Kagaku* (Science) 9: 113–114.

———. 1961. Edward Divers sensei to Haga Tamemasa sensei. *Kagaku* (Chemistry) 16: 782–786.

Shibusawa, Eiichi. 1932. Takamatsu Hakushi. In *Takamatsu Toyokichi den* (1932): 93-100.

Shinoda, Kōzō. 1944. Meiji Bunkashi jō no Iseki Kōbu Daigakkō. In *Meiji bunka no shin kenkyū*, ed. Osatake Takeki, pp. 201–232. Tokyo: Ajia Shobō.

Shiohara, Matasaku, ed. 1926. *Takamine Hakushi*. Tokyo: Shiohara Matasaku.

Shiokawa, Hisao. 1977. R. W. Atkinson: On his Life and Drinking Water Analysis (in Japanese). *Kagakushi* no. 6: 20–24.

———. 1978. On the Analysis of the Drinking Water in the First Half of Meiji Era (in Japanese). *Kagakushi* no. 7: 20–26.

Shunpo Kō Tsuishōkai, ed. 1940. *Itō Hirobumi den*. 3 vols. Tokyo: Shunpo Kō Tsuishōkai.

Sidar, Jean Wilson. 1976. *George Hammell Cook: A Life in Agriculture and Geology*. New Brunswick, NJ: Rutgers University Press.

Simões, Ana, Ana Carneiro, and Maria Paula Diogo, eds. 2003. *Travels of Learning: A Geography of Science in Europe*. Dordrecht, Boston, and London: Kluwer Academic Publishers.

Simon, Josep. 2011. *Communicating Physics: The Production, Circulation and Appropriation of Ganot's Textbooks in France and England, 1851–1887.* London: Pickering & Chatto.

———. 2012. Cross-National Education and the Making of Science, Technology, and Medicine. *Hist. Sci.* 50: 251–256.

Smith, Crosbie, and Jon Agar. 1998. Introduction: Making Space for Science. In *Making Space for Science: Territorial Themes in the Shaping of Knowledge*, ed. Crosbie Smith and Jon Agar, pp. 1–23. Basingstoke: Macmillan; New York: St. Martin's Press.

Smith, Robert H. 1897. The University of Japan. *The Engineer* 84: 582.

Smith, Watson, and T. Takamatsu. 1881a. On Phenylnaphthalene. *JCST* 39: 546–551.

———, and T. Takamatsu. 1881b. Sulphonic Acids derived from Isodinaphthyl (ββ- Dinaphtyl). *JCST* 39: 551–554.

———, and T. Takamatsu. 1882. On Pentathionic Acid (Part II). *JCST* 41: 162–167.

Stranges, Anthony N. 1993. Synthetic Fuel Production in Prewar and World War II Japan. *Ann. Sci.* 50: 229–65.

Sugai, Jun-ichi, and Tanaka Minoru, eds. 1970. *Nkgt 13: Butsuri kagaku.* Tokyo: Dai-ichi Hōki.

Sugawara, Kunika. 1978. Meiji chūki no bunseki kagaku: 1890 nendai. *Bunseki* 1978 no. 3: 186–190.

Sugimoto, Isao. 1962. *Kinsei jitsugakushi no kenkyū: Edo jidai chūki ni okeru kagaku gijutsugaku no seisei.* Tokyo: Yoshikawa Kōbunkan.

———. 1967. Jiseiteki kagaku bunka: Jitsugaku no kōryū. In *Kagakushi: Taikei nihonshi sōsho 19*, ed. Sugimoto Isao, pp. 148–204. Tokyo: Yamakawa Shuppansha.

Sugiyama, Shinya. 1981. Glover & Co.: A British Merchant in Nagasaki, 1861–1870. In *Bakumatsu and Meiji: Studies in Japan's Economic and Social History*, ed. Ian Nish, pp. 1–16. London: London School of Economics and Political Science.

Suito, Eiji. 1983. Academic Achievement and Career of Dr. Shinkichi Horida (in Japanese). *Kagakushi* 1983, no. 1: 19–32.

Swale, Alistair. 2000. *The Political Thought of Mori Arinori: A Study in Meiji Conservatism.* Richmond, Surrey: Japan Library.

Takamatsu, Tōru. 2001. Kagaku kōgyō. In *Sangyō gijutsushi*, eds. Nakaoka Tetsurō, Suzuki Jun, Tsutsumi Ichirō, and Miyachi Masato, pp. 284–320. Tokyo: Yamakawa Shuppansha.

Takamatsu, Toyokichi. 1878. *On Japanese Pigments.* Tokyo: The Department of Science, Tokio Daigaku.

———. 1882. Jinkō seiran no setsu. *Tkk* 3: 21–33.

———. 1883a. Tokyo Kagakukai Dai-5 nenkai no shukuji ni kau. *Tkk* 4: 48–54.

———. 1883b. Kaki junkai no hōkoku. *Tkk* 4: 107–117.

———. 1891. Ranten seizō shiken hōkoku. *Tkk* 12: 297–317.

———. 1892. Ranten seizō shiken hōkoku, Dai 2 kai. *Tkk* 13: 287–300.

———. 1893. Ranten seizō shiken hōkoku, Dai 3 kai. *Tkk* 14: 337–351.

———. 1895a. *Senshokuhō: Senshokuhō Sōron, Sen-i, Seirenhō, hyōhakuhō, baisenzai, shikiso, shinsenhō, nassenhō, haishokuhō.* Tokyo: Maruzen.

———. 1895b. Ranten seizō ni tuki hōkoku. *Tkk* 16: 1–12.

Takamatsu, Toyokichi, and Watson Smith. 1880. On Pentathionic Acid. *JCST* 37: 592–608.

Takamatsu hakushi shukuga denki kankōkai, ed. 1932. *Kōgaku hakushi Takamatsu Toyokichi den.* Tokyo: Kagaku Kōgyō Jihōsha.

Tamamushi, Bun-ichi. 1978. Kaimen kagaku eno michi: Katayama Masao kyōju seitan 100 nen ni chinande. *Kagakushi* no. 8: 1–6.

Tanaka, Akira, and Seiji Takata, eds. 1993. *"Beiō kairan Jikki" no gakusai-teki kenkyū.* Sapporo: Hokkaidō Daigaku Tosho Kankōkai.

Tanaka, Fujimaro, 1875. *Riji kōtei.* 15 vols. Tokyo: Monbusho.

Tanaka, Minoru. 1964. Kagaku. In *Meiji-zen nihon butsuri kagakushi*, ed. Nihon Gakushiin, pp. 281–432. Tokyo: Maruzen.

———. 1975. *Nihon no kagaku to Shibata Yūji.* Tokyo: Dai Nihon Tosho.

Tanaka, Minoru, and Yamazaki Toshio. 1964. *Nkgt 21: Kagaku gijutsu.* Tokyo: Dai-ichi Hōki.

Tanaka, Yoshio. 1961a. Takamatsu Toyokichi sensei. *Kagaku* (Chemistry) 16: 706–708.

———. 1961b. Kawakita Michitada sensei. *Kagaku* (Chemistry) 16: 1070–1072.

Tarbell, Dean Stanley, and Ann Tracy Tarbell. 1986. The Johns Hopkins University, Ira Remsen and Organic Chemistry, 1876–1913. In *Essays on the History of Organic Chemistry in the United States*, eds. D. Stanley Tarbell and Ann Tracy Tarbell, pp. 24–39. Nashville, Tenn.: Folio Publishers.

Taylor, Georgette. 2008. Marking out a Disciplinary Common Ground: The Role of Chemical Pedagogy in Establishing the Doctrine of Affinity at the Heart of British Chemistry. *Ann. Sci.* 65: 465–486.

Terasaki, Masao. 2000a (1979). *Nihon ni okeru daigaku jichi seido no seiritsu.* Zōho-ban edition. Tokyo: Hyōronsha.

———. 2000b (1973–1974). "Kōza sei" no rekisiteki kenkyū josetsu. In Terasaki, *Nihon ni okeru Daigaku Jichi Seido no Seiritsu*, Zōho-ban edition, pp. 371–411. Tokyo: Hyōronsha.

Thackray, Arnold, Jeffrey L. Sturchio, P. Thomas Carroll, and Robert Bud. 1985. *Chemistry in America, 1876–1976: Historical Indicators.* Dordrecht, Boston, and Lancaster: D. Reidel.

Thorpe, T. E. 1873. *Quantitative Chemical Analysis.* London: Longman, Green.

Thorpe, T. E., and M. M. Pattison Muir. 1874. *Qualitative Chemical Analysis and Laboratory Practice.* London: Longman, Green.

Tōhoku Daigaku, ed. 1960. *Tōhoku Daigaku gojūnenshi.* 2 vols. Sendai: Tōhoku Daigaku.

Tōhoku Daigaku hyakunenshi henshū iinkai, ed. 2003–2010. *Tōhoku Daigaku hyakunenshi*. 11 vols. Sendai: Tōhoku Daigaku Shuppannkai.

Tōhoku kagaku dōsōkai, ed. 1992. *Tōhoku kagaku dōsōkaihō: Kagaku kyōshitsu sōritsu hachijusshūnen kinengō*. Sendai: Tōhoku Kagaku Dōsōkai.

Tokyo Daigaku Daigakuin Rigakukei Kenkyūka Rigakubu Kagaku Kyōshitsu Zasshikai, ed. 2007. *Tokyo Daigaku Rigakubu Kagaku Kyōshitsu no ayumi: Sōsetsuki kara sengo madeno kiroku*. Tokyo: Tokyo Daigaku Daigakuin Rigakukei Kenkyūka Rigakubu Kagaku Kyōshitsu Zasshikai.

Tokyo Daigaku hyakunenshi hensan iinkai, ed. 1984–1987. *Tokyo Daigaku hyakunenshi*. 10 vols. Tokyo: Tokyo Daigaku Shuppankai.

Tokyo Daigakushi shiryō kenkyūkai, ed. 1993–1994. *Tokyo Daigaku nenpō*. 6 vols. Tokyo: Tokyo Daigaku Shuppankai.

Tokyo Daigakushi Shiryōshitsu. 2005. *Watanabe Hiromoto shiryō mokuroku*. Tokyo: Tokyo Daigakushi Shiryōshitsu.

Tokyo Kōgyō Daigaku. 1940. *Tokyo Kōgyō Daigaku rokujunenshi*. Tokyo: Tokyo Kogyo Daigaku.

Tokyo Teikoku Daigaku. 1932. *Tokyo Teikoku Daigaku gojunenshi*. 2 vols. Tokyo: Tokyo Teikoku Daigaku.

Tomory, Leslie. 2011. Gaslight, Distillation, and the Industrial Revolution. *Hist. Sci.* 49: 395–424.

Tomozawa, Junji. 1965. R. W. Atkinson's Study of Sake Brewing (in Japanese). *Kk* 75: 114–123.

Tsuda, Sakae. 1956. Kyōikusha to shiteno Ikeda sensei. In *Tsuiokuroku* (1956): 153–155.

Tsuji, Shinji. 1882. Waga kuni kagaku no kigen. *Tkk* 3: 64–74.

Tsuji, Tetsuo. 1973. *Nihon no kagaku shisō: Sono jiritsu eno mosaku*. Tokyo: Chūō Kōronsha.

Tsukahara, Togo. 1993. *Affinity and Shinwa Ryoku: Introduction of Western Chemical Concepts in Early Nineteenth-Century Japan*. Amsterdam: J. C. Gieben.

———. 1994. Elimination of Qi by Chemical Specification: Shift of Understanding of Western Theory of Matter in Japan. *Historia Scientiarum* 4: 1–23.

Tsukahara, Tokumichi. 1978a. *Meiji kagaku no kaitakusha*. Tokyo: Sanseidō.

———, ed. 1978b. *Kuhara Mitsuru shokanshū*. Tsuyama: Tsuyama Yōgaku Shiryōkan.

Uchida, Takane, Oki Hisaya, Sakan Fujio, Isa Kimio, and Nakata Ryuji. 1990. William Elliot Griffis's Lecture Notes on Chemistry. In *Foreign Employees in Nineteenth- Century Japan*, eds. Edward R. Beauchamp and Akira Iriye, pp. 247–258. Boulder, San Francisco, and London: Westview Press.

Uchida, Yoshichika. 1999. Daigaku ni okeru senmon kyōiku to kōza sei. *Kenchiku zasshi. Kenchiku nenpō* 1990: 28.

Ueda, Shōzaburō. 1954. Kyōiku dantaishi. In *Kyōiku bunkashi taikei*, vol. 5, ed. Ishiyama Shūhei et. al., pp. 219–257. Tokyo: Kaneko Shobō.

Ueda, Toyokichi, ed. 1925. *Waguneru den*. Tokyo: Hakurankai Shuppan Kyōkai.

Uemura, Shoji. 2008. Salaries of oyatoi (Japan's Foreign Employees) in Early Meiji (in Japanese). *Ryūtsū Kagaku Daigaku ronshū: Ryūtsū keiei hen* 21: 1–24.

Ueno, Masuzō. 1968. Robāto Wiriamu Atokinson. In *Oyatoi gaikokujin 3: Shizen kagaku*, ed. Ueno Masuzō, pp. 113–136. Tokyo: Kashima Shuppankai.

Umeda, Otogorō, ed. 1938. *Waguneru sensei tsuikaishū*. Tokyo: Ko Waguneru hakushi kinen jigyōkai.

Umetani, Noboru. 1971. *The Role of Foreign Employees in the Meiji Era in Japan*. Tokyo: Institute of Developing Economies.

UNESCO Higashi Ajia Bunka Kenkyū Sentā, ed. 1975. *Shiryō oyatoi gaikokujin* Tokyo: Shōgakkan.

University of London. 1912. *The Historical Record (1836–1912): Being a Supplement to the Calendar Completed to September 1912*, First Issue. London: University of London Press.

Ure, Andrew. 1856. *A Dictionary of Arts, Manufactures, and Mines; contained A Clear Exposition of Their Principles and Practice*. 2 vols. New York: D. Appleton.

Usselman, Melvyn C., Christina Reinhart, Kelly Foulser, and Alan J. Rocke. 2005. Restaging Liebig: A Study in the Replication of Experiments. *Ann. Sci.* 62: 1–55.

Veysey, Laurence R. 1965. *The Emergence of the American University*. Chicago and London: University of Chicago Press.

Wada, Masanori. 2011. The Role of the Ministry of Public Works in Meiji Japan in Designing Engineering Education: Reconsidering the Foundation of the Imperial College of Engineering (in Japanese). *Kk* 50: 86–96.

———. 2012. Chemistry and Practical Education at the Imperial College of Engineering in Tokyo, 1873–1886 (in Japanese). *Kagakushi* 39: 55–78.

Wanklyn, J. Alfred, and Ernest Theophron Chapman. 1868. *Water Analysis: A Practical Treatise on the Examination of Potable Water*. London: Trübner.

Warner, Debora J. 2008. Ira Remsen, Saccharin, and the Linear Model. *Ambix* 55: 50–61.

Warwick, Andrew. 2003. *Masters of Theory: Cambridge and the Rise of Mathematical Physics*. Chicago: University of Chicago Press.

Watanabe, Hiromoto. 1886. Ri ka ryōgaku no kōeki o minkan ni hanpu senkoto o tsutomu beshi. *Tkk* 7: 27–32.

Watanabe, Masao. 1973. F. F. Jewett: Chemist and Educator (in Japanese). *Kk* Series II 12: 223–228.

Watt, Carey A., and Michael Mann, eds. 2011. *Civilizing Missions in Colonial and Postcolonial South Asia: From Improvement to Development*. London and New York: Anthem Press.

Wetzler, Peter. 1998. *Hirohito and War: Imperial Tradition and Military Decision Making in Prewar Japan.* Honolulu: University of Hawai'i Press.

Williamson, Alexander William. 1849. *Development of Difference the Basis of Unity: Introductory Lecture to the Courses of the Faculty of Arts and Laws, delivered in University College, London, October 16, 1849.* London: Taylor, Walton, and Maberly.

———. 1851–1854. Suggestions for the Dynamics of Chemistry Derived from the Theory of Etherification. *Proceedings of the Royal Institution of Great Britain* 1: 90–94.

———. 1865. *Chemistry for Students.* Oxford: Clarendon Press.

———. 1870. *A Plea for Pure Science: Being the Inaugural Lecture at the Opening of the Faculty of Science, in University College, London, October 4th, 1870.* London: Printed by Taylor and Francis.

Wittner, David G. 2008. *Technology and the Culture of Progress in Meiji Japan.* Abingdon, Oxon, and New York: Routledge.

Wright, David. 2000. *Translating Science: The Transmission of Western Chemistry into Late Imperial China, 1840–1900.* Leiden: Brill.

Yamamoto, Yumiyo. 1997. Viscount Inoue: Father of the Japanese Railways. *TASJ* 4th series 12: 89–104.

Yamamoto, Shoji. 1987. William Elliot Griffis as an Interpreter of the Meiji Society: The First American Envoy of New Japan. Ph.D. dissertation, State University of New York at Buffalo.

Yamamuro, Shin-ichi, and Nakanome Tōru, eds. 1999–2009. *Meiroku Zasshi.* 3 vols. Tokyo: Iwanami Shoten.

Yamazaki, Eiichi, and Yamada Nobuo. 1920. Jiasutāze no kagaku (dai 1 pō): Hannō sokudoron no kenchi yori bakugatō no kōzō kettei. *Tkk* 41: 621–690.

Yasue, Masakazu. 1983. On the Academic Achievement of N. Nagai and the Vicissitude of his Academic Career (in Japanese). *Kagakushi* 1983 no. 1: 31–8.

———. 1986. Reexamination of the Chemical Achievements of Nagajosi Nagai: A Critique of Traditional Writings about Him (in Japanese). *Kagakushi* 1986 no. 4:159–168.

Yorita, Michio. 1984. Waguneru to kōgyō gijutsu kyōiku. In *Meiji Ishin Kyōikushi*, ed. Inoue Hisao, pp. 233–253. Tokyo: Yoshikawa Kōbunkan.

Yoshida, Hikorokurō. 1883. Chemistry of Lacquer (Urushi). Part I. *JCST* 43: 472–486.

———. 1885. Chemical Examination of the Constituents of Camphor Oil. *JCST* 47: 779-801.

Yoshida, Tadashi. 1974. The Rangaku of Shizuki Tadao: The Introduction of Western Science in Tokugawa Japan. Ph.D. dissertation, Princeton University.

———. 1984. Tenbō: Rangakushi. *Kk* 23: 73–80.

Yoshihara, H. Kenji. 2000. Ogawa's Discovery of Nipponium and Its Re-evaluation. *Historia Scientiarum* 9: 257–269.

———. 2003. Masataka Ogawa and his Students at Tohoku University: Research of Platinum Group Elements and X-ray Spectroscopy (in Japanese). *Kagakushi* 30: 69–83.

———. 2008. Nipponium as a new Element (Z + 75) Separated by the Japanese chemist, Masataka Ogawa: A Scientific and Science Historical Reevaluation. *Proceedings of the Japan Academy*, series B 84: 232–245.

Yoshiie, Sadao. 1992. David Murray, Superintendent of Educational Affairs in Japan: His Views on Education and his Influences in Japan and in the United States. Ph.D. dissertation, State University of New York at Buffalo.

———. 1998. *Nihonkoku gakkan Deibiddo Marē: Sono shōgai to gyōseki.* Tokyo: Tamagawa Daigaku Shuppanbu.

Index

Note: Page numbers in italics denote concepts with explanations while those with 'n' denote page numbers taken from notes.